Advancing Maths for AQA
STATISTICS 2

Gill Buque, Roger Lugsdin and Roger Williamson

Series editors
Roger Williamson   Sam Boardman   Graham Eaton
Ted Graham   Keith Parramore

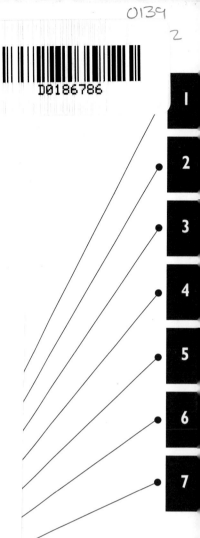

heinemann.co.uk

✓ Free online support
✓ Useful weblinks
✓ 24 hour online ordering

01865 888058

Heinemann
Inspiring generations

Heinemann is an imprint of Pearson Education Limited,
a company incorporated in England and Wales, having
its registered office at Edinburgh Gate, Harlow, Essex,
CM20 2JE. Registered company number: 872828

Heinemann is a registered trademark of Pearson Education Limited

Text © Gill Buque, Roger Lugsdin and Roger Williamson 2000, 2005
Complete work © Heinemann Educational Publishers 2005

First published 2005

10 09
10 9 8 7 6 5 4

British Library Cataloguing in Publication Data is available from the British
Library on request.

ISBN 978 0 435513 39 9

Edited by Alex Sharpe, Standard Eight Limited
Typeset and illustrated by Tech-Set Limited, Gateshead, Tyne & Wear
Original illustrations © Harcourt Education Limited, 2004
Cover design by Miller, Craig and Cocking Ltd
Printed in China (CTPS/04)

Acknowledgements
The publishers and authors acknowledge the work of the writers Ray Atkin,
John Berry, Derek Collins, Tim Cross, Ted Graham, Phil Rawlins, Tom Roper,
Rob Summerson, Nigel Price, Frank Chorlton and Andy Martin of the *AEB
Mathematics for AQA A-Level Series*, from which some exercises and examples
have been taken.

The publishers' and authors' thanks are due to AQA for permission to
reproduce questions from past examination papers.

The answers have been provided by the authors and are not the responsibility
of the examining board.

Every effort has been made to contact copyright holders of material reproduced
in this book. Any omissions will be rectified in subsequent printings if notice is
given to the publishers.

About this book

This book is one in a series of textbooks designed to provide you with exceptional preparation for AQA's 2004 Mathematics Specification. The series authors are all senior members of the examining team and have prepared the textbooks specifically to support you in studying this course.

Finding your way around

The following are there to help you find your way around when you are studying and revising:

- **edge marks** (shown on the front page) – these help you to get to the right chapter quickly;

- **contents list** – this identifies the individual sections dealing with key syllabus concepts so that you can go straight to the areas that you are looking for;

- **index** – a number in bold type indicates where to find the main entry for that topic.

Key points

Key points are not only summarised at the end of each chapter but are also boxed and highlighted within the text like this:

> The standard deviation of a discrete random variable, X, is defined by:
>
> $$\text{Standard deviation of } X = \sqrt{\text{Var}(X)} = \sqrt{\text{E}(X^2) - \mu^2}$$

Exercises and exam questions

Worked examples and carefully graded questions familiarise you with the syllabus and bring you up to exam standard. Each book contains:

- Worked examples and Worked exam questions to show you how to tackle typical questions; Examiner's tips will also provide guidance;

- Graded exercises, gradually increasing in difficulty up to exam-level questions, which are marked by an [A];

- Test-yourself sections for each chapter so that you can check your understanding of the key aspects of that chapter and identify any sections that you should review;

- Answers to the questions are included at the end of the book.

Appendix

Answers

Index

CHAPTER I

Discrete probability distributions

Learning objectives

After studying this chapter, you should be able to:

- understand what is meant by a discrete random variable
- understand what is meant by expectation, mean, standard deviation and variance of a discrete random variable
- find the mean, standard deviation and variance of a discrete random variable
- find the mean, standard deviation and variance of a simple function of a discrete random variable.

1.1 Introduction

In S1 you met one discrete probability distribution.
This is called the binomial probability distribution.

It is possible to list the values of a discrete variable, together with their associated probabilities.

You can do this in two ways:

- You can represent the probability distribution in the form of a table.
- You can define a simple function which allocates probabilities to each value of the discrete variable.

Worked example 1.1

Find the probability distribution for the discrete random variable X, where X is the number of heads obtained when a fair coin is tossed three times.

Solution

You should remember, from your work in S1 on the binomial distribution, that $X \sim B(3, \frac{1}{2})$, with X taking the values 0, 1, 2, 3.

Probabilities can be allocated to these discrete values of X by using the simple formula:

$$P(X = x) = \binom{3}{x}\left(\frac{1}{2}\right)^x\left(\frac{1}{2}\right)^{3-x} = \binom{3}{x}\left(\frac{1}{2}\right)^3 \quad \text{for } x = 0, 1, 2, 3.$$

The probabilities are allocated as follows:

$$P(X = 0) = \binom{3}{0}\left(\frac{1}{2}\right)^3 = \frac{1}{8} \qquad P(X = 2) = \binom{3}{2}\left(\frac{1}{2}\right)^3 = \frac{3}{8}$$

$$P(X = 1) = \binom{3}{1}\left(\frac{1}{2}\right)^3 = \frac{3}{8} \qquad P(X = 3) = \binom{3}{3}\left(\frac{1}{2}\right)^3 = \frac{1}{8}$$

The probability distribution can also be defined in the form of the table:

x	0	1	2	3
$P(X = x)$	$\frac{1}{8}$	$\frac{3}{8}$	$\frac{3}{8}$	$\frac{1}{8}$

You could also determine this probability distribution using a tree diagram to enable you to find the probabilities allocated to each value of the discrete variable X.

> In the examination you may be expected to determine the probability distribution by considering a simple problem.

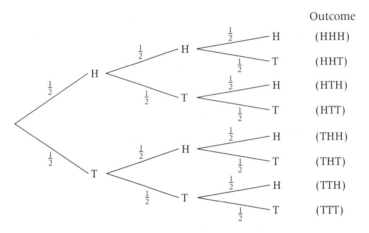

		Outcome
	H	(HHH)
	T	(HHT)
	H	(HTH)
	T	(HTT)
	H	(THH)
	T	(THT)
	H	(TTH)
	T	(TTT)

Also, since X is a discrete **random** variable, the sum of the probabilities equals one.

This is usually written as $\displaystyle\sum_{\text{all } x} P(X = x) = 1$.

> A random variable is a variable whose value is (within limits) determined by chance.

Worked example 1.2

A discrete random variable, X, can take the values 0, 2, 3, 4, 5 and 6, with probabilities allocated in the following way:

$$P(X = x) = \begin{cases} \dfrac{1}{32}(x + 1) & \text{for } x = 0, 3, 4, 5 \\[2mm] \dfrac{1}{16}x & \text{for } x = 2, 6 \end{cases}$$

> This is sometimes referred to as the probability density function.

Write down the probability distribution in the form of a table.

Solution

$$P(X = 0) = \frac{1}{32} \qquad P(X = 4) = \frac{5}{32}$$

$$P(X = 3) = \frac{4}{32} = \frac{1}{8} \qquad P(X = 5) = \frac{6}{32} = \frac{3}{16}$$

$$P(X = 2) = \frac{2}{16} = \frac{1}{8} \qquad P(X = 6) = \frac{6}{16} = \frac{3}{8}$$

x	0	2	3	4	5	6
P(X = x)	$\frac{1}{32}$	$\frac{1}{8}$	$\frac{1}{8}$	$\frac{5}{32}$	$\frac{3}{16}$	$\frac{3}{8}$

> Notice that the probabilities add up to one.
> $$\sum_{\text{all } x} P(X = x) = 1$$

> X is a **random** variable.

Worked example 1.3

A discrete random variable, X, has a probability density function defined by $P(X = x) = kx^2$ for $x = 1, 2, 3, 4$ and 5.

(a) Write down the probability distribution in the form of a table.

(b) Find the value of k.

(c) Find $P(X < 3)$.

Solution

(a)

x	1	2	3	4	5
$P(X = x)$	k	$4k$	$9k$	$16k$	$25k$

(b) Since X is a random variable, $\displaystyle\sum_{\text{all } x} P(X = x) = 1$

$$\therefore \quad k + 4k + 9k + 16k + 25k = 1$$
$$55k = 1$$
$$k = \frac{1}{55}$$

(c) $P(X < 3) = P(X = 1 \text{ or } X = 2) = k + 4k = 5k$
$$= \frac{1}{11}$$

Worked example 1.4

A discrete random variable, R, has a probability distribution defined by

r	1	2	4	8
P(R = r)	q	0.1	$2q$	0.3

Find:

(a) the value of q,

(b) $P(R \geqslant 2)$.

Solution

(a) $\sum_{\text{all } r} P(R = r) = 1$ gives $q + 0.1 + 2q + 0.3 = 1$

$$3q + 0.4 = 1$$
$$q = 0.2$$

(b) $P(R \geqslant 2) = P(R = 2 \text{ or } R = 4 \text{ or } R = 8)$
$$= 0.1 + 0.4 + 0.3$$
$$= 0.8$$

> You could have used
> $$P(R \geqslant 2) = 1 - P(R < 2)$$
> $$= 1 - 0.2$$
> $$= 0.8$$

Worked example 1.5

A blue tetrahedral die has the numbers 1, 3, 5 and 7 on its four faces.
A red tetrahedral die has the numbers 2, 4, 6 and 8 on its four faces.
The two dice are thrown and the number that each die lands on is recorded.

The discrete variable, T, is the total obtained when the number that each of these two **fair** dice lands on are added together.

(a) Determine the probability distribution for T, in the form of a table.

(b) Show that T is a discrete **random** variable.

(c) Find $P(7 \leqslant T < 13)$.

Solution

(a) The possible outcomes can be represented as follows:

+	2	4	6	8
1	3	5	7	9
3	5	7	9	11
5	7	9	11	13
7	9	11	13	15

This gives the probability distribution defined by the table:

t	3	5	7	9	11	13	15
$P(T = t)$	$\frac{1}{16}$	$\frac{1}{8}$	$\frac{3}{16}$	$\frac{1}{4}$	$\frac{3}{16}$	$\frac{1}{8}$	$\frac{1}{16}$

(b) $\sum_{\text{all } t} P(T = t) = \frac{1}{16} + \frac{1}{8} + \frac{3}{16} + \frac{1}{4} + \frac{3}{16} + \frac{1}{8} + \frac{1}{16} = 1$
$\therefore$ T is a **random** variable.

(c) $P(7 \leqslant T < 13) = P(T = 7 \text{ or } T = 9 \text{ or } T = 11)$
$$= \frac{5}{8}$$

EXERCISE 1A

1 A discrete random variable, X, has a probability density function defined by

$$P(X = x) = ax^3 \quad \text{for } x = 1, 2, 3 \text{ and } 4.$$

(a) Represent the probability distribution for X in the form of a table.

(b) Calculate the value of the constant a.

(c) Find $P(X < 3)$.

2 A discrete random variable, R, has a probability distribution defined by

r	0	4	8	12
$P(R = r)$	0.05	s	$4s$	0.25

(a) Calculate the value of the constant s.

(b) Find $P(R \geqslant 8)$.

3 A discrete random variable, Y, has probability distribution defined by

y	-2	-1	1	2	3
$P(Y = y)$	0.1	c	c	$4c$	0.3

(a) Calculate the value of the constant c.

(b) Find:
  (i) $P(Y < 0)$,
  (ii) $P(-1 < Y \leqslant 2)$.

1.2 The expectation of a discrete random variable

> The expected value of a discrete random variable, X, is usually denoted by $E(X)$, where
> $$E(X) = \sum_{\text{all } x} x \, P(X = x).$$

Worked example 1.6

A discrete random variable, X, has the probability distribution

x	0	2	8	10
$P(X = x)$	0.4	0.3	0.2	0.1

Find the value of $E(X)$.

Solution

$$E(X) = \sum_{\text{all } x} x \, P(X = x) = (0 \times 0.4) + (2 \times 0.3) + (8 \times 0.2) + (10 \times 0.1)$$
$$= 3.2$$

Random variables are usually denoted by an uppercase letter such as X.
Particular values of a random variable are usually denoted by lowercase letters such as x.

Alternatively, we could simply extend the table given in the question:

x	0	2	8	10	
P(X = x)	0.4	0.3	0.2	0.1	$\sum P(X=x) = 1$
$x\,\mathbf{P}(X=x)$	0	0.6	1.6	1.0	$\sum x\,P(X=x) = 3.2$

$E(X)$ is the mean score we would expect to obtain if samples were repeatedly taken from this distribution.

1.3 The mean of a discrete random variable

> The mean of a discrete random variable is defined to be $E(X)$.

$\therefore$ The mean of X is $E(X) = \sum_{\text{all } x} x\,P(X = x)$.

We often write the mean of X as $\mu = E(X)$.

Worked example 1.7

The probability distribution of a discrete random variable, R, is shown in the following table:

r	1	2	4	8
P(R = r)	0.2	0.3	0.35	0.15

Find:

(a) the mean of R, (b) $P(R < 5)$.

Solution

(a) $E(R) = \sum_{\text{all } r} r\,P(R = r)$
$= (1 \times 0.2) + (2 \times 0.3) + (4 \times 0.35) + (8 \times 0.15)$
$= 3.4$

You could write this as $\mu = 3.4$.

(b) $P(R < 5) = P(R = 1 \text{ or } R = 2 \text{ or } R = 4)$
$= 0.2 + 0.3 + 0.35$
$= 0.85$

You could have used
$$P(R < 5) = 1 - P(R \geqslant 5)$$
$$= 1 - 0.15$$
$$= 0.85$$

Worked example 1.8

A discrete random variable, Y, has a probability density function $P(Y = y) = ay^{\frac{1}{2}}$ for $y = 1, 4, 9, 16$ and 25.

(a) Write, in terms of a, the probability distribution in the form of a table.

(b) Find:

(i) the value of the constant a,

(ii) the mean of Y,

(iii) $P(Y > 12)$.

Solution

(a)

y	1	4	9	16	25
$P(Y = y)$	a	$2a$	$3a$	$4a$	$5a$

(b) (i) $\displaystyle\sum_{\text{all } y} P(Y = y) = 1$

$$15a = 1$$
$$a = \frac{1}{15}$$

(ii) $\displaystyle \mu = E(Y) = \sum_{\text{all } y} y\, P(Y = y)$

$$= (1 \times a) + (4 \times 2a) + (9 \times 3a) + (16 \times 4a) + (25 \times 5a)$$
$$= 225a$$
$$= 225 \times \frac{1}{15}$$
$$= 15$$

(iii) $P(Y > 12) = P(Y = 16 \text{ or } Y = 25)$

$$= \frac{4}{15} + \frac{5}{15}$$
$$= \frac{3}{5}$$

EXERCISE 1B

1 A discrete random variable, X, has a probability distribution defined by:

x	2	4	6	12
$P(X = x)$	0.25	0.15	0.20	0.40

Calculate the mean of X.

2 A bag contains five blue counters and four red counters. Two counters are drawn at random from the bag, without replacement.

(a) Find, in the form of a table, the probability distribution for the discrete random variable X, the number of blue counters drawn from the bag.

(b) Calculate $E(X)$.

3 The probability that Sam scores in any particular game of football is 0.2, independent of all other games of football in which she plays.

(a) Find, in the form of a table, the probability distribution of the discrete random variable S, the number of games in which Sam scores over the first three games of the season.

(b) Find the mean of S.

(c) Write down the probability that Sam scores in at most one of these games.

4 The discrete random variable R has probability distribution defined by $P(R = r) = kr^2$, for $r = 1, 3, 4$ and 8, and k is a constant.

(a) Find the value of the constant k.

(b) Calculate the mean of R.

1.4 The expectation of g(X)

> The expectation of $g(X)$, denoted by $E[g(X)]$, where $g(X)$ is any function of a discrete random variable X having a probability distribution determined by $P(X = x)$ is defined to be:
>
> $$E[g(X)] = \sum_{\text{all } x} g(x)P(X = x).$$

Worked example 1.9

A discrete random variable, X, has the following probability distribution:

x	2	4	6	8
$P(X = x)$	0.1	0.2	0.5	0.2

(a) Calculate $E(X)$.

(b) (i) Write, in the form of a table, the probability distribution for $g(X) = 4X$.

(ii) Calculate $E(4X)$.

(c) Show that $E(4X) = 4 \times E(X)$.

Solution

(a) $E(X) = \sum_{\text{all } x} x\, P(X = x) = (2 \times 0.1) + (4 \times 0.2) + (6 \times 0.5)$
$$+ (8 \times 0.2)$$
$$= 5.6$$

(b) (i)

$4x$	8	16	24	32
$P(X = x)$	0.1	0.2	0.5	0.2

(ii) $E(4X) = \sum_{\text{all } x} 4x\, P(X = x)$
$$= (8 \times 0.1) + (16 \times 0.2) + (24 \times 0.5) + (32 \times 0.2)$$
$$= 22.4$$

(c) $4 \times E(X) = 4 \times 5.6$
$$= 22.4$$
$$= E(4X)$$

In general $E(aX) = aE(X)$.

Worked example 1.10

The discrete random variable, X, has probability distribution defined by:

x	1	2	3	4	5
$P(X = x)$	0.05	0.45	0.15	0.25	0.1

(a) Calculate $E(X)$.

(b) (i) Tabulate the distribution for $Y = 4X + 6$.

(ii) Hence find the mean value of the perimeter of rectangles with sides of length X and $X + 3$.

(c) Show that $E(Y) = 4E(X) + 6$.

Solution

(a) $E(X) = \sum_{\text{all } x} x\, P(X = x)$

$\quad\quad = (1 \times 0.05) + (2 \times 0.45) + (3 \times 0.15) + (4 \times 0.25)$
$\quad\quad\quad\quad + (5 \times 0.1)$

$\quad\quad = 2.9$

(b) (i) Distribution for $Y = 4X + 6$

y	10	14	18	22	26
$P(Y = y)$	0.05	0.45	0.15	0.25	0.1

(ii) Perimeter $= 4X + 6 = Y$

$E(Y) = \sum_{\text{all } y} y\, P(Y = y)$

$\quad\quad = (10 \times 0.05) + (14 \times 0.45) + (18 \times 0.15)$
$\quad\quad\quad\quad + (22 \times 0.25) + (26 \times 0.1)$

$\quad\quad = 17.6$

(c) $4E(X) + 6 = 4 \times 2.9 + 6$
$\quad\quad\quad\quad = 11.6 + 6$
$\quad\quad\quad\quad = 17.6$
$\quad\quad\quad\quad = E(Y)$

In general $E(aX + b) = aE(X) + b$, where a and b are constants.

Worked example 1.11

A discrete random variable, R, has probability distribution defined by:

r	0	1	2	4	5
$P(R = r)$	0.2	0.15	0.35	0.2	0.1

Find:

(a) $E(R)$,

(b) $E(R^2)$.

Solution

(a) $E(R) = \sum_{\text{all } r} r\, P(R = r)$

$= (0 \times 0.2) + (1 \times 0.15) + (2 \times 0.35) + (4 \times 0.2)$
$\qquad + (5 \times 0.1)$

$= 2.15$

(b) $E(R^2) = \sum_{\text{all } r} r^2\, P(R = r)$

$= (0^2 \times 0.2) + (1^2 \times 0.15) + (2^2 \times 0.35)$
$\qquad + (4^2 \times 0.2) + (5^2 \times 0.1)$

$= (0 \times 0.2) + (1 \times 0.15) + (4 \times 0.35) + (16 \times 0.2)$
$\qquad + (25 \times 0.1)$

$= 7.25$

> Note that it is the value of R that is squared and not the probability.

To try to avoid the common mistake made in examinations of squaring the probability instead of the value of R, you could use a table as demonstrated below.

r^2	0	1	4	16	25	
P(R = r)	0.2	0.15	0.35	0.2	0.1	$\sum_{\text{all } r} P(R = r) = 1$
$r^2\, \mathbf{P(R = r)}$	0	0.15	1.40	3.20	2.5	$\sum_{\text{all } r} r^2\, P(R = r) = 7.25$

Thus giving $E(R^2) = 7.25$.

> Note that $E(R^2) \neq [E(R)]^2$.

Worked example 1.12

A discrete random variable, X, has a probability density function defined by:

x	1	2	4	5
P(X = x)	0.2	0.1	0.5	0.2

(a) Find:

    (i) $E(X)$,   (ii) $E(X^{-1})$.

(b) Hence find:

    (i) $E(10X - 4)$,   (ii) $E(10X^{-1})$.

Solution

(a) (i) $E(X) = \sum_{\text{all } x} x\, P(X = x)$

$= (1 \times 0.2) + (2 \times 0.1) + (4 \times 0.5) + (5 \times 0.2)$
$= 3.4$

    (ii) $E(X^{-1}) = E\!\left(\dfrac{1}{X}\right) = \sum_{\text{all } x} \dfrac{1}{x} P(X = x)$

$= (1 \times 0.2) + \left(\tfrac{1}{2} \times 0.1\right) + \left(\tfrac{1}{4} \times 0.5\right)$
$\qquad + \left(\tfrac{1}{5} \times 0.2\right)$

$= 0.2 + 0.05 + 0.125 + 0.04$

$= 0.415$

> Note that $E\!\left(\dfrac{1}{X}\right) \neq \dfrac{1}{E(X)}$.

(b) (i) $\mathrm{E}(10X - 4) = 10\mathrm{E}(X) - 4$
$$= 10 \times 3.4 - 4$$
$$= 30$$

(ii) $\mathrm{E}(10X^{-1}) = 10\mathrm{E}(X^{-1})$
$$= 10 \times 0.415$$
$$= 4.15$$

Use of the result $\mathrm{E}(aX) = a\mathrm{E}(X)$, where a is a constant.

EXERCISE 1C

1 A discrete random variable, Y, has a probability distribution defined by:

y	-2	-1	1	2	5	10
$\mathbf{P}(Y = y)$	0.1	0.2	0.3	0.2	0.1	0.1

(a) Write down, in the form of a table, the probability distribution for Y^{-2}.

(b) Hence calculate:
    (i) $\mathrm{E}(Y^{-2})$,   **(ii)** $\mathrm{E}(100Y^{-2})$.

2 A discrete random variable, X, has the following probability distribution:

x	0	1	2	5	10
$\mathbf{P}(X = x)$	0.1	0.2	0.3	0.3	0.1

(a) Calculate:
    (i) $\mathrm{E}(X)$,   **(ii)** $\mathrm{E}(X^2)$.

(b) Hence find:
    (i) $\mathrm{E}(5X)$,   **(ii)** $\mathrm{E}(25X^2)$.

3 A discrete random variable, R, has a probability density function defined by:

r	1	2	5	10	50
$\mathbf{P}(R = r)$	0.3	0.2	0.1	0.25	0.15

(a) Calculate $\mathrm{E}(R^{-1})$.

(b) Squares have sides of length $\dfrac{5}{R}$. Find the value for the mean perimeters of the squares.

1.5 The variance and standard deviation of a discrete random variable

The variance of a discrete random variable, X, is defined by:
$$\mathrm{Var}(X) = \mathrm{E}(X - \mathrm{E}(X))^2$$
$$= \mathrm{E}(X^2) - [\mathrm{E}(X)]^2$$
$$= \mathrm{E}(X^2) - \mu^2$$

The variance of X is usually denoted by:
$$\mathrm{Var}(X) = \sigma^2.$$

> The standard deviation of a discrete random variable, X, is defined by:
>
> $$\text{Standard deviation of } X = \sqrt{\text{Var}(X)} = \sqrt{E(X^2) - \mu^2}$$

The standard deviation is usually denoted by σ.

Worked example 1.13

A discrete random variable, X, has the following probability distribution:

x	1	2	3	4	5
$P(X = x)$	0.1	0.3	0.2	0.3	0.1

(a) Find:

    (i) $E(X)$,

    (ii) $E(X^2)$.

(b) Hence calculate the value of $\text{Var}(X)$.

Solution

(a) (i) $E(X) = \sum_{\text{all } x} x\, P(X = x)$

$$= (1 \times 0.1) + (2 \times 0.3) + (3 \times 0.2) + (4 \times 0.3) + (5 \times 0.1)$$

$$= 3.0$$

    (ii) $E(X^2) = \sum_{\text{all } x} x^2\, P(X = x)$

$$= (1^2 \times 0.1) + (2^2 \times 0.3) + (3^2 \times 0.2) + (4^2 \times 0.3) + (5^2 \times 0.1)$$

$$= 10.4$$

(b) $\text{Var}(X) = E(X^2) - \mu^2 = 10.4 - (3.0)^2$

$$= 1.4$$

Worked example 1.14

In a 'pay and display' car park, motorists use an automatic machine to purchase tickets. The price £X, of the ticket depends on the length of time the motorist intends to leave the car in the car park.

The following probability distribution provides a suitable model for the discrete random variable X:

x	$P(X = x)$
0.8	0.25
1.4	0.55
2.0	0.08
2.8	p

(a) Find the value of p.

(b) Calculate the mean and the standard deviation of X.

(c) A few motorists park but do not purchase a ticket. It is decided to modify the probability distribution to include these motorists by allocating a small probability to the outcome $X = 0$. Will the standard deviation of this modified probability distribution be greater than, the same as or smaller than the value calculated in (b)?

Solution

(a) $\displaystyle\sum_{\text{all } x} P(X = x) = 1$

$\therefore \quad 0.25 + 0.55 + 0.08 + p = 1$

$$p = 1 - 0.88$$
$$= 0.12$$

(b) $\displaystyle \mu = E(X) = \sum_{\text{all } x} x\, P(X = x)$

$$= (0.8 \times 0.25) + (1.4 \times 0.55) + (2.0 \times 0.08)$$
$$+ (2.8 \times 0.12)$$
$$= 1.466$$

$\displaystyle E(X^2) = \sum_{\text{all } x} x^2\, P(X = x)$

$$= (0.8^2 \times 0.25) + (1.4^2 \times 0.55) + (2.0^2 \times 0.08)$$
$$+ (2.8^2 \times 0.12)$$
$$= 2.4988$$

$$\sigma^2 = \text{Var}(X) = E(X^2) - \mu^2 = 2.4988 - (1.466)^2$$
$$= 0.349\,644$$

$\sigma = $ standard deviation $= \sqrt{0.349\,644} = 0.591$

> Standard deviation, $\sigma = \sqrt{\text{variance}}$.

(c) If $X = 0$ is included in the probability distribution it will be more spread out and so the standard deviation will be increased.

EXERCISE 1D

1 A discrete random variable, X, has a probability distribution defined by:

x	0	1	2	3
$P(X = x)$	0.4	0.3	0.2	0.1

Find the mean, variance and standard deviation of X.

2 A discrete random variable, Y, has a probability distribution defined by:

y	0	1	4	10
$P(Y = y)$	0.2	0.5	0.2	0.1

Find the mean, variance and standard deviation of Y.

3 A discrete random variable, T, has probability density function defined by:

t	0	1	2	3	4
$P(T = t)$	0.2	0.4	0.2	0.1	p

(a) Find the value of p.

(b) Find the mean, variance and standard deviation of T.

4 Members of a public library may borrow up to four books at any one time. The number of books borrowed by a member on each visit to the library is a random variable, X, with the following probability distribution:

x	$P(X = x)$
0	0.24
1	0.12
2	0.20
3	0.28
4	0.16

Find the mean and standard deviation of X.

1.6 The variance of a simple function of a discrete random variable

> $\text{Var}(a) = 0$,
> $\text{Var}(aX) = a^2\,\text{Var}(X)$,
> $\text{Var}(aX + b) = a^2\,\text{Var}(X)$,
>
> where a and b are constants.

The standard deviation is obtained by finding the square root of the variance.

Worked example 1.15

A discrete random variable, R, has the following probability distribution:

r	1	2	3	5
$P(R = r)$	0.2	0.4	0.1	0.3

(a) Find:

    (i) $E(R)$,

    (ii) $E(R^2)$.

(b) Hence find the variance and standard deviation of R.

(c) (i) Find, in the form of a table, the probability distribution for $3R + 2$.

    (ii) Hence find $\text{Var}(3R + 2)$.

(d) Show that $\text{Var}(3R + 2) = 9\text{Var}(R)$.

Solution

(a) (i)
$$E(R) = \sum_{\text{all } r} r\, P(R = r)$$
$$= (1 \times 0.2) + (2 \times 0.4) + (3 \times 0.1) + (5 \times 0.3)$$
$$= 0.2 + 0.8 + 0.3 + 1.5$$
$$= 2.8$$

(ii)
$$E(R^2) = \sum_{\text{all } r} r^2\, P(R = r)$$
$$= (1^2 \times 0.2) + (2^2 \times 0.4) + (3^2 \times 0.1)$$
$$\qquad + (5^2 \times 0.3)$$
$$= 0.2 + 1.6 + 0.9 + 7.5$$
$$= 10.2$$

(b)
$$\sigma^2 = \text{Var}(R) = E(R^2) - [E(R)]^2$$
$$= 10.2 - (2.8)^2$$
$$= 2.36$$

Standard deviation $= \sigma = \sqrt{2.36} = 1.54$.

(c) (i)

$3R + 2$	5	8	11	17
$P(R = r)$	0.2	0.4	0.1	0.3

(ii) $E(3R + 2) = 3E(R) + 2 = 3 \times 2.8 + 2 = 10.4$

$$\text{Var}(3R + 2) = \sum_{\text{all } r} (3r + 2)^2\, P(R = r) - [E(3R + 2)]^2$$
$$= (5^2 \times 0.2) + (8^2 \times 0.4) + (11^2 \times 0.1)$$
$$\qquad + (17^2 \times 0.3) - (10.4)^2$$
$$= 129.4 - 108.16$$
$$= 21.24$$

> This demonstrates the use of the general result
> $\text{Var}(aR + b) = a^2 \times \text{Var}(R)$ using $a = 3$ and $b = 2$.

(d) $9 \times \text{Var}(R) = 9 \times 2.36 = 21.24 = \text{Var}(3R + 2)$

Worked example 1.16

A discrete random variable, Y, has a probability distribution defined by:

y	-4	-2	0	2	4
$P(Y = y)$	0.1	0.1	0.2	0.4	0.2

(a) Calculate: **(i)** $E(Y)$, **(ii)** $\text{Var}(Y)$.

(b) Hence find: **(i)** $E(5Y - 3)$, **(ii)** $\text{Var}(5Y + 4)$.

Solution

(a) (i)
$$E(Y) = \sum_{\text{all } y} y\, P(Y = y)$$
$$= (-4 \times 0.1) + (-2 \times 0.1) + (0 \times 0.2)$$
$$\qquad + (2 \times 0.4) + (4 \times 0.2)$$
$$= -0.4 - 0.2 + 0 + 0.8 + 0.8$$
$$= 1.0$$

(ii) $\mathrm{E}(Y^2) = \sum_{\text{all } y} y^2\, \mathrm{P}(Y = y)$

$= (16 \times 0.1) + (4 \times 0.1) + (0 \times 0.2)$
$\qquad + (4 \times 0.4) + (16 \times 0.2)$
$= 1.6 + 0.4 + 0 + 1.6 + 3.2$
$= 6.8$

$\mathrm{Var}(Y) = \mathrm{E}(Y^2) - [\mathrm{E}(Y)]^2$
$\quad = 6.8 - (1.0)^2$
$\quad = 5.8$

> Remember to square the y-values and not the probabilities.

> Don't forget to subtract μ^2.

(b) (i) $\mathrm{E}(5Y - 3) = 5\mathrm{E}(Y) - 3$
$\qquad\qquad = 5 \times 1.0 - 3$
$\qquad\qquad = 2$

> Use of the general result $\mathrm{E}(aY + b) = a\mathrm{E}(Y) + b$, where a and b are constants.

(ii) $\mathrm{Var}(5Y + 4) = 5^2\,\mathrm{Var}(Y)$
$\qquad\qquad = 25 \times 5.8$
$\qquad\qquad = 145$

> Use of the general result $\mathrm{Var}(aY + b) = a^2\,\mathrm{Var}(Y)$, where a and b are constants.

Worked example 1.17

A discrete random variable, X, has a probability distribution defined by:

x	1	4	5	10
$\mathbf{P}(X = x)$	0.05	0.12	0.48	0.35

(a) Find the mean and standard deviation of X.

(b) Find $\mathrm{E}(Y)$ and $\mathrm{Var}(Y)$ by tabulating the distribution for $Y = X + \dfrac{1}{X}$.

(c) Hence find the mean and variance of the **perimeter**, P, of rectangles having sides of length $X + \dfrac{5}{X}$ and $4X$.

Solution

(a) $\mathrm{E}(X) = \sum_{\text{all } x} x\, \mathrm{P}(X = x)$

$= (1 \times 0.05) + (4 \times 0.12) + (5 \times 0.48) + (10 \times 0.35)$
$= 0.05 + 0.48 + 2.4 + 3.5$
$= 6.43$

$\mathrm{E}(X^2) = \sum_{\text{all } x} x^2\, \mathrm{P}(X = x)$

$= (1^2 \times 0.05) + (4^2 \times 0.12) + (5^2 \times 0.48)$
$\qquad + (10^2 \times 0.35)$
$= 0.05 + 1.92 + 12.0 + 35.0$
$= 48.97$

$\therefore \quad \mathrm{Var}(X) = \mathrm{E}(X^2) - [\mathrm{E}(X)]^2$
$\qquad\qquad = 48.97 - (6.43)^2$
$\qquad\qquad = 7.6251$

$\therefore \quad$ standard deviation $= \sqrt{7.6251}$
$\qquad\qquad\qquad\qquad = 2.76$

(b) $Y = X + \dfrac{1}{X}$

y	2	4.25	5.2	10.1
P($Y = y$)	0.05	0.12	0.48	0.35

$$E(Y) = \sum_{\text{all } y} y\,P(Y = y)$$
$$= (2 \times 0.05) + (4.25 \times 0.12) + (5.2 \times 0.48)$$
$$+ (10.1 \times 0.35)$$
$$= 0.10 + 0.51 + 2.496 + 3.535$$
$$= 6.641$$
$$E(Y^2) = \sum_{\text{all } y} y^2\,P(Y = y)$$
$$= (2^2 \times 0.05) + (4.25^2 \times 0.12) + (5.2^2 \times 0.48)$$
$$+ (10.1^2 \times 0.35)$$
$$= 0.2 + 2.1675 + 12.9792 + 35.7035$$
$$= 0.2 + 2.1675 + 12.9792 + 35.7035$$
$$= 51.0502$$
$$\therefore \quad \text{Var}(Y) = 51.0502 - 6.641^2$$
$$= 6.947\,319$$

(c) Perimeter of rectangle, P, is given by:

$$P = 2\left(X + \frac{5}{X}\right) + 2 \times 4X$$

$$P = 10X + \frac{10}{X} = 10\left(X + \frac{1}{X}\right) = 10Y$$

$$E(P) = E(10Y)$$
$$= 10E(Y)$$
$$= 10 \times 6.641$$
$$= 66.41$$
$$\text{Var}(P) = \text{Var}(10Y)$$
$$= 10^2 \times \text{Var}(Y)$$
$$= 100 \times 6.947\,319$$
$$= 695$$

EXERCISE 1E

1 A discrete random variable, X, has a probability density function defined by:

x	1	2	3	4
P($X = x$)	0.15	0.2	0.25	0.4

(a) Calculate the mean and variance of X.

(b) Hence find:
   (i) $E(3X)$,
   (ii) $\text{Var}(3X + 4)$.

2 A discrete random variable, Y, has a probability distribution defined by:

y	2	4	6	8
$P(Y = y)$	0.4	0.3	0.2	0.1

(a) Calculate the mean and variance of Y.

(b) Hence find:
   (i) $E(10Y + 4)$,
   (ii) $Var(10Y + 4)$.

3 A bag contains six red counters and four blue counters. Two counters are drawn at random from the bag, without replacement. The number of red counters, R, drawn from the bag is a discrete random variable.

(a) Show that the probability distribution for R can be defined by:

r	0	1	2
$P(R = r)$	$\frac{2}{15}$	$\frac{8}{15}$	$\frac{1}{3}$

(b) Calculate the mean and variance of R.

(c) Hence find $Var(5R + 2)$.

4 A biased coin is such that when the coin is tossed in the air, the probability of obtaining a head is 0.4.

The coin is tossed in the air 10 times.
The number of heads obtained, Y, is a discrete random variable.

(a) Describe the probability distribution for the discrete random variable Y.

(b) Hence write down the value of the mean and standard deviation of Y.

(c) Find:
   (i) $E(10Y + 8)$,   (ii) $Var(10Y - 6)$.

5 The price, £Y, of tickets for the theatre depends on the position of the seat in the theatre. The following probability distribution provides a suitable model for the discrete random variable Y:

y	$P(Y = y)$
25	0.32
27.50	0.48
32	0.15
40	0.05

(a) Calculate the mean and standard deviation of Y.

(b) For a gala performance evening, the price of theatre tickets is doubled. Find the mean and standard deviation of the price of the tickets for this gala performance.

(c) For a special performance of a popular play, 5% of the cheapest tickets are given, free of charge, to the students of the local School of Dramatic Art.
  (i) If there are 2000 seats in the theatre, how many tickets were allocated to the students?
  (ii) Calculate the total takings for this special performance, assuming that all the tickets are sold.
  (iii) State, with a reason, what you consider will be the effect, of the free tickets, on the mean and standard deviation of the price of theatre tickets for this special performance.

MIXED EXERCISE

1 A discrete random variable, X, has a probability distribution defined by:

x	0	1	2	3
$P(X=x)$	0.05	0.15	0.35	k

(a) Find the value of k.

(b) Find the mean, variance and standard deviation of X.

2 A discrete random variable, T, has a probability distribution defined by:

t	3	5	7	9
$P(T=t)$	0.24	0.35	0.23	0.18

Find the mean, variance and standard deviation of T.

3 A regular customer at a small clothes shop observes that the number of customers, X, in the shop when she enters has the following probability distribution.

Number of customers, x	$P(X=x)$
0	0.15
1	0.34
2	0.27
3	0.14
4	0.10

Find the mean and standard deviation of X.

4 A discrete random variable, R, has the following probability distribution:

r	1	2	4	5
$P(R=r)$	0.1	0.3	0.2	0.4

(a) Write down, in the form of a table, the probability distribution of $\frac{1}{R}$.

(b) Calculate the values of $E(R^{-1})$ and $E(R^{-2})$.

(c) Hence find the variance and standard deviation of $\frac{10}{R}$.

5 A discrete random variable, X, has a probability distribution defined by:

$$f(x) = 0.05x, \text{ for } x = 2, 4, 6 \text{ and } 8.$$

(a) Write down, in the form of a table, the probability distribution of X.

(b) Calculate the mean and variance of X.

Rectangles have sides of length $3X + 2$ and $2X + 3$.

(c) Calculate the mean and standard deviation of the **perimeter** of the rectangles.

6 A company produces blue carpet material. The length of material (in metres) required to meet each order is a discrete random variable, X, with the following probability distribution:

x	$P(X = x)$
50	0.50
60	0.08
70	0.04
80	0.05
90	0.08
100	0.25

(a) Find the mean and standard deviation of X.

(b) What is the probability that on a day during which exactly two orders are placed the total length of material ordered is 120 m?   [A]

7 Prospective recruits to a large retailing organisation undergo a medical examination. As part of the examination, their heights are measured by a nurse and recorded to the nearest 2 mm. The final digit of the recorded height may be modelled by a discrete random variable, X, with the following probability distribution:

x	0	2	4	6	8
$P(X = x)$	0.2	0.2	0.2	0.2	0.2

(a) Find the mean and standard deviation of X.

(b) A new nurse recorded the heights to the nearest 5 mm. Construct an appropriate probability distribution for the final digit of the recorded height.   [A]

8 Applicants for a sales job are tested on their knowledge of consumer protection legislation. The test consists of five multiple choice questions. The number of correct answers, X, follows the probability distribution:

x	0	1	2	3	4	5
$P(X = x)$	0.60	0.04	0.07	0.10	0.09	0.10

(a) Find the mean and standard deviation of X.

A group of production staff, who had no knowledge of the subject, guessed all of the answers. The probability of each answer being correct was 0.25. The discrete random variable, Y, represents the distribution of the number of correct answers for this group.

(b) (i) Name the probability distribution which could provide a suitable model for Y.

(ii) Determine the mean and standard deviation of Y.

(c) Compare and comment briefly on the results of your calculations in (a) and (b)(ii). [A]

9 A village inn offers bed and breakfast and has four bedrooms available for customers. The number of bedrooms occupied each night may be modelled by the random variable X, with the following probability distribution:

x	$P(X = x)$
0	0.22
1	0.25
2	0.20
3	0.14
4	0.19

(a) Calculate:
(i) the mean of X,
(ii) $E(X^2)$,
(iii) the standard deviation of X.

(b) Find the probability that the number of occupied bedrooms:
(i) is less than 3,
(ii) exceeds the mean. [A]

10 Marian belongs to the Handchester Building Society. She frequently visits her local branch to pay instalments on her mortgage. The number of people queuing to be served when she enters the branch may be modelled by the random variable X, with the following probability distribution:

x	$P(X = x)$
0	0.12
1	0.33
2	0.27
3	0.18
4	0.07
5	0.03

(a) Find the probability that when she enters the branch there are two or more people queuing to be served.

(b) Find:

   (i) the mean of X,

   (ii) $E(X^2)$,

   (iii) the standard deviation of X.   [A]

11 Doctor Patel works at a health centre. On every Monday morning she has appointments with five patients.

   (a) Experience suggests that on each Monday the number of these patients who had not booked an appointment with her during the previous year may be modelled by the random variable X.

   The probability distribution of X is given in the following table:

x	$P(X = x)$
0	0.20
1	0.34
2	0.31
3	0.10
4	0.04
5	0.01

   Find:

   (i) the mean of X,

   (ii) $E(X^2)$,

   (iii) the standard deviation of X.

   (b) Of all patients registered with Doctor Patel, 40% have not booked an appointment with her during the previous year. Five patients, registered with Doctor Patel, were selected at random.

   (i) Name the distribution which could provide a model for the number of patients, Y, in the sample, who had not booked an appointment with her during the previous year.

   (ii) Write down the mean and standard deviation of Y.   [A]

12 A discrete random variable R is such that $E(R) = 3$ and $\text{Var}(R) = 1$.

   (a) Determine the mean and variance of $5(R - 1)$.

   The probability distribution of R is:

$$P(R = r) = \begin{cases} \dfrac{r}{10} & r = 1, 2, 3, 4 \\ 0 & \text{otherwise.} \end{cases}$$

   (b) Calculate the mean and variance of $12R^{-1}$.   [A]

13 The probability distribution for the number of vehicles, V, involved in each minor accident on a particular stretch of road can be modelled as follows:

v	1	2	3	4	5
$P(V = v)$	0.15	0.45	0.20	0.15	0.05

(a) Show that $E(V) = 2.5$ and $Var(V) = 1.15$.

(b) The total cost, £C, of removing all the damaged vehicles following a minor accident is given by $C = 30V + 25$. Determine the mean and variance of C.

(c) The total repair cost, £R, for all the vehicles involved in a minor accident is given by $R = 40V^2 + 15V + 50$. Determine the value of $E(R)$. [A]

14 The discrete random variable T is such that $E(T) = 5$ and $Var(T) = 25$.

(a) Rectangles have sides of length $2T$ and $(T + 5)$. Determine the mean and variance of the **perimeter** of the rectangles.

(b) (i) Show that $E(T^2) = 50$.
 (ii) Hence determine the mean of the **area** of the rectangles. [A]

15 The probability distribution for the number, R, of unwrapped sweets in a tin is given in the following table:

r	0	1	2	3	4
$P(R = r)$	0.1	0.2	0.4	0.2	0.1

(a) Show that:
 (i) $E(R) = 2$,
 (ii) $Var(R) = 1.2$.

(b) The number, P, of partially wrapped sweets in a tin is given by $P = 3R + 4$. Find values for $E(P)$ and $Var(P)$.

(c) The total number of wrapped sweets in a tin is 200. Sweets are either correctly wrapped, partially wrapped or unwrapped.
 (i) Express C, the number of correctly wrapped sweets in a tin, in terms of R.
 (ii) Hence find the mean and variance of C. [A]

16 The probability distribution for a discrete random variable R is tabulated below.

r	1	2	3	4	5
$P(R = r)$	0.1	0.2	0.4	0.2	0.1

(a) Given that $E(R) = 3$ and $Var(R) = 1.2$, find the mean and variance of $5(2R - 1)$.

(b) **(i)** Write down the probability distribution for $\dfrac{60}{R}$.

(ii) Show that $E\left(\dfrac{60}{R}\right) = 24.2$.

(iii) Given that $E\left(\dfrac{3600}{R^2}\right) = 759.4$, determine the variance of $\dfrac{60}{R}$. [A]

Key point summary

1 For a discrete random variable X,   *p5*

$$E(X) = \sum_{\text{all } x} x\, P(X = x).$$

2 If $g(X)$ is any function of a discrete random   *p8*
variable X,

$$E[g(X)] = \sum_{\text{all } x} g(x) P(X = x).$$

3 Mean $= \mu = E(X)$ *p6*

4 Variance $= \sigma^2 = \text{Var}(X) = E[X - E(X)]^2$ *p11, 12*
$$= E(X^2) - [E(X)]^2$$
$$= E(X^2) - \mu^2$$
Standard deviation $= \sigma = \sqrt{\text{Var}(X)}$

5 For a discrete random variable X, and constants a *p14*
and b,

$E(a) = a$ $\text{Var}(a) = 0$
$E(aX) = aE(X)$ $\text{Var}(aX) = a^2\,\text{Var}(X)$
$E(aX + b) = aE(X) + b$ $\text{Var}(aX + b) = a^2\,\text{Var}(X)$

6 You will also be expected to know and to be   *S1, chapter 4*
able to use the following results relating to
the binomial probability distribution.

If $X \sim B(n, p)$ then: $E(X) = \mu = np$
$$\text{Var}(X) = \sigma^2 = np(1 - p),$$

where $P(X = x) = \dbinom{n}{x} p^x (1 - p)^{n-x}$.

Test yourself	**What to review**

1 A discrete random variable, X, has a probability density function defined by $P(X = x) = kx^3$, for $x = 1, 2, 3$ and 4.

Section 1.2

  (a) Write down the probability distribution in the form of a table.

  (b) Find the value of k.

  (c) Find $P(X > 2)$.

2 A discrete random variable, R, has a probability density function defined by:

Sections 1.3 and 1.5

r	0.25	0.5	1.0	2.5
$P(R = r)$	0.4	0.3	0.2	0.1

Find the mean, variance and standard deviation of R.

3 A discrete random variable, X, has a probability distribution defined by:

Sections 1.4 and 1.6

x	1	2	3	4	5
$P(X = x)$	0.2	0.3	0.1	0.25	0.15

  (a) Find $E(X)$ and $Var(X)$,

  (b) Hence find the values of $E(5X)$ and $Var(5X)$.

4 A discrete random variable, X, has a probability distribution defined by:

Sections 1.4 and 1.6

x	1	2	5	10
$P(X = x)$	0.4	0.2	0.3	0.1

By tabulating the probability distribution for $Y = \dfrac{10}{X}$, find:

  (i) $E(10X^{-1})$ and **(ii)** $Var(10X^{-1})$.

5 A discrete random variable, Y, has a probability density function defined by:

y	0	2	4	8
$P(Y = y)$	0.05	0.35	0.5	0.1

  (a) Calculate the mean, standard deviation and variance of Y.   *Sections 1.3 and 1.5*

  (b) Hence find: **(i)** $E(10Y + 7)$ and **(ii)** $Var(5Y + 7)$.   *Sections 1.4 and 1.6*

6 A discrete random variable X has the following probability distribution:

Sections 1.4 and 1.6

x	1	2	5	10	20
$P(X = x)$	0.4	0.3	0.15	0.1	0.05

  (a) (i) Calculate $E(X^{-1})$ and $E(X^{-2})$,
     (ii) Hence find $Var(X^{-1})$.

  (b) Use the answers to **(a)** to find $E(6X^{-1})$ and $Var(6X^{-1})$.

  (c) Calculate the value of $E(100X^{-2})$.

1 (a)

x	1	2	3	4
$P(X = x)$	k	$8k$	$27k$	$64k$

; **(b)** $k = 0.01$; **(c)** 0.91.

2 $E(R) = 0.7$, $Var(R) = 0.435$, standard deviation $= 0.660$.

3 (a) $E(X) = 2.85$, $Var(X) = 1.93$; **(b)** $E(5X) = 14.25$, $Var(5X) = 48.2$.

4 (i) $E(10X^{-1}) = 5.7$, **(ii)** $Var(10X^{-1}) = 13.8$.

5 (a) mean $= 3.5$, variance $= 3.55$, standard deviation $= 1.88$;
(b) (i) 42, **(ii)** 88.75.

6 (a) (i) $E(X^{-1}) = 0.5925$ and $E(X^{-2}) = 0.482$, **(ii)** $Var(X^{-1}) = 0.131$;
(b) $E(6X^{-1}) = 3.555$ and $Var(6X^{-1}) = 4.72$; **(c)** 48.2.

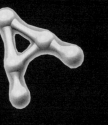

CHAPTER 2

The Poisson distribution

2

Learning objectives

After studying this chapter, you should be able to:
■ recognise circumstances where a Poisson distribution will provide a suitable model
■ use tables of the Poisson distribution
■ calculate Poisson probabilities.

2.1 Introduction

The Poisson distribution arises when events occur independently, at random at a constant average rate.

For example the number of cars passing a point, per minute, on a quiet stretch of motorway might be modelled by a Poisson distribution. The number of telephone calls arriving at a switchboard over a 5-minute interval might also be modelled by a Poisson distribution. As for the binomial distribution (which you met in MS01), only discrete, whole number outcomes are possible (0, 1, 2, 3, 4 …). However, unlike the binomial distribution there is no upper limit to the possible number of outcomes.

A constant average rate does *not* mean that the same number of cars pass the point in each minute.

Other examples where the Poisson distribution might provide a suitable model are:

● the number of faults in a metre of dressmaking material
● the number of accidents per month on a particular stretch of motorway
● the number of daisies in a square metre of lawn.

2.2 The Poisson distribution

The French mathematician Siméon-Denis Poisson showed that if events occur, in a given interval, independently at random at a constant average rate λ (that is if they follow a Poisson distribution), the probability that exactly r events will occur in a particular interval is:

$$\frac{e^{-\lambda}\lambda^r}{r!}$$

This distribution is sometimes denoted Po(λ).

2.3 Tables of the Poisson distribution

It is often unnecessary to use the Poisson formula because tables of the cumulative Poisson distribution are available. As with the binomial tables these tabulate the probability of '*r* or fewer' events occurring. An extract is shown below.

Table 2 Cumulative Poisson Distribution

The tabulated value is $P(X \leqslant x)$, where X has a Poisson distribution with mean λ.

λ / x	0.1	0.2	0.3	0.4	0.5	0.6	0.7	0.8	0.9	0.10	1.2	1.4	1.6	1.8	λ / x
0	0.9048	0.8187	0.7408	0.6703	0.6065	0.5488	0.4966	0.4493	0.4066	0.3679	0.3012	0.2466	0.2019	0.1653	0
1	0.9953	0.9825	0.9631	0.9384	0.9098	0.8781	0.8442	0.8088	0.7725	0.7358	0.6626	0.5918	0.5249	0.4628	1
2	0.9998	0.9989	0.9964	0.9921	0.9856	0.9769	0.9659	0.9526	0.9371	0.9197	0.8795	0.8335	0.7834	0.7306	2
3	1.000	0.9999	0.9997	0.9992	0.9982	0.9966	0.9942	0.9909	0.9865	0.9810	0.9662	0.9463	0.9212	0.8913	3
4		1.000	1.000	0.9999	0.9998	0.9996	0.9992	0.9986	0.9977	0.9963	0.9923	0.9857	0.9763	0.9636	4
5				1.000	1.000	1.000	0.9999	0.9998	0.9997	0.9994	0.9985	0.9968	0.9940	0.9896	5
6							1.000	1.000	1.000	0.9999	0.9997	0.9994	0.9987	0.9974	6
7										1.000	1.000	0.9999	0.9997	0.9994	7
8												1.000	1.000	0.9999	8
9														1.000	9

For example, for a Poisson distribution with mean 0.9 the probability of two or fewer events occurring is 0.9371.

Worked example 2.1

On average eight vehicles pass a point on a free-flowing motorway in a 10-second interval. Find the probability that in a particular 10-second interval the number of cars passing this point is:

> We can use the Poisson distribution here since we know the vehicles arrive at a constant average rate and that the traffic is free-flowing so that vehicles arrive independently and at random during this time interval.

(a) seven or fewer

(b) 12 or more

(c) fewer than nine

(d) more than eight

(e) exactly nine

(f) between seven and 10, inclusive.

Solution

(a) $P(X \leqslant 7) = 0.4530$
$= 0.453$

> Here we use the $\lambda = 8.0$ column of the Poisson distribution tables.

(b) $P(X \geqslant 12) = 1 - P(X \leqslant 11)$
$= 1 - 0.8881$
$= 0.1119$
$= 0.112$

11 | 12 13

(c) $P(X < 9) = P(X \leqslant 8)$
$= 0.5925$

$7\ \ 8\ \boxed{9\ \ 10}$

(d) $P(X > 8) = 1 - P(X \leqslant 8)$
$= 1 - 0.5925$
$= 0.4075$

$\overline{7\ \ 8}\,|\,9\ \ 10$

2

(e) $P(X = 9) = P(X \leqslant 9) - P(X \leqslant 8)$
$= 0.7166 - 0.5925$
$= 0.1241$
$= 0.124$

$7\ \ 8\,\boxed{9}\,10\ \ 11$

(f) $P(7 \leqslant X \leqslant 10) = P(X \leqslant 10) - P(X \leqslant 6)$
$= 0.8159 - 0.3134$
$= 0.5025$

$5\ \ 6\,\boxed{7\ \ 8\ \ 9\ \ 10}\,11\ \ 12$

EXERCISE 2A

1 The number of telephone calls arriving at a switchboard follow a Poisson distribution with mean 6 per 10-minute interval. Find the probability that the number of calls arriving in a particular 10-minute interval is:

   (a) eight or fewer,

   (b) more than three,

   (c) fewer than six,

   (d) between four and seven inclusive,

   (e) exactly seven.

2 The number of customers arriving at a supermarket checkout in a 10-minute interval may be modelled by a Poisson distribution with mean 4. Find the probability that the number of customers arriving in a specific 10-minute interval is:

   (a) two or fewer,

   (b) fewer than seven,

   (c) exactly three,

   (d) between three and seven inclusive,

   (e) five or more.

3 People arrive at a ticket office independently, at random, at an average rate of 12 per half-hour. Find the probability that the number of people arriving in a particular half-hour interval is:

   (a) exactly 10,

   (b) fewer than eight,

   (c) more than 16,

   (d) between nine and 15 inclusive,

   (e) 14 or fewer.

4 The number of births announced in the personal column of a local weekly newspaper may be modelled by a Poisson distribution with mean 9.5. Find the probability that the number of births announced in a particular week will be:

   (a) fewer than five,

   (b) between six and 12 inclusive,

   (c) eight or more,

   (d) exactly 11,

   (e) more than 11.

5 The number of people joining a checkout queue at a supermarket may be modelled by a Poisson distribution with a mean of 1.8 per minute. Find the probability that in a particular minute the number of people joining the queue is:

   (a) one or fewer,

   (b) exactly three.

6 The number of letters of complaint received by a department store follows a Poisson distribution with mean 6.5 per day. Find the probability that on a particular day,

   (a) 7 or fewer letters of complaint are received,

   (b) exactly 7 letters of complaint are received.

7 The weekly number of ladders sold by a small DIY shop can be modelled by a Poisson distribution with mean 1.4. Find the probability that in a particular week the shop will sell:

   (a) 2 or fewer ladders,

   (b) exactly 4 ladders,

   (c) 2 or more ladders.

Worked example 2.2

As part of a feasibility study into introducing tolls on a motorway it is estimated that the number of cars arriving at a toll-booth site could be modelled by a Poisson distribution with mean 3.4 per 10-second interval. It is recommended that k toll-booths be installed where the number of cars arriving is less than or equal to k in at least 85% of 10-second intervals. Find k.

Solution

This question is answered by reading down the column $\lambda = 3.4$ until we find the first probability greater than 0.85. In this case we find that the probability of five or fewer cars arriving in a 10-second interval is 0.8705. Hence the required value of k is 5.

EXERCISE 2B

1 The number of customers arriving at an office selling tickets for a festival, may be modelled by a Poisson distribution with mean 1.2 per 2-minute interval. Find the number of arrivals which will not be exceeded in at least 90% of 2-minute intervals.

2 A garage offering a quick change service for exhausts finds that the demand for exhausts to fit a Metro may be modelled by a Poisson distribution with mean 8 per day. Find the demand which will not be exceeded on at least:

   (a) 95% of days,

   (b) 99% of days,

   (c) 99.8% of days.

3 A small newsagent finds that weekday demand for *The Independent* follows a Poisson distribution with mean 12. How many *Independents* should the newsagent stock if the demand is to be satisfied on at least:

   (a) 90% of days,

   (b) 99% of days.

4 Demand for an item in a warehouse may be modelled by a Poisson distribution with mean 14 per day. The warehouse can only be stocked at the beginning of each day. How many items should the stock be made up to in order to ensure that demand can be met on:

   (a) 95% of days,

   (b) 99% of days,

   (c) 99.5% of days.

2.4 Calculating Poisson probabilities

Tables of the Poisson distribution, although quick to use, may not always be sufficiently detailed to enable us to obtain the probabilities required. In this case we will need to use the fact that if the random variable, R, follows a Poisson distribution with mean λ then:

$$P(R = r) = e^{-\lambda}\frac{\lambda^r}{r!} \text{ for } r = 0, 1, 2, 3\ldots$$

Remember $0! = 1$.

Worked example 2.3

The number of calls arriving at a switchboard may be modelled by a Poisson distribution with mean 4.3 per 10-minute interval.

Find the probability that in a 10-minute interval:

(a) no calls arrive,

(b) exactly one call arrives,

(c) exactly two calls arrive,

(d) exactly three calls arrive,

(e) between one and three calls, inclusive, arrive,

(f) more than two calls arrive.

Solution

(a) $P(R = 0) = \dfrac{e^{-4.3}4.3^0}{0!} = 0.013\,568\,6 = 0.0136$

(b) $P(R = 1) = \dfrac{e^{-4.3}4.3^1}{1!} = 0.058\,344\,8 = 0.0583$

(c) $P(R = 2) = \dfrac{e^{-4.3}4.3^2}{2!} = 0.125\,441\,3 = 0.125$

(d) $P(R = 3) = \dfrac{e^{-4.3}4.3^3}{3!} = 0.179\,799\,2 = 0.180$

(e) $P(1 \le R \le 3) = P(R = 1) + P(R = 2) + P(R = 3)$
$= 0.0583\,45 + 0.125\,44 + 0.179\,80$
$= 0.364$

(f) $P(R > 2) = 1 - P(R \le 2)$
$= 1 - [P(R = 0) + P(R = 1) + P(R = 2)]$
$= 1 - [0.013\,569 + 0.058\,345 + 0.125\,44]$
$= 1 - 0.197\,35$
$= 0.803$

> These probabilities can be obtained directly from some calculators. This is acceptable in the examinations.

> 0 | 1 2 3 | 4
> Using the answers above correct to five significant figures in order to give the final answer correct to three significant figures.

> 0 1 2 | 3 4 5
> Using the above results correct to five significant figures in order to give the final answer correct to three significant figures.

EXERCISE 2C

1 In a Poisson distribution the mean rate of occurrence of an event is 5.3 per hour. Find the probability that in a given hour:

(a) none of these events happen,

(b) exactly one of these events happens,

(c) exactly two of these events happen,

(d) exactly three of these events happen,

(e) more than three of these events happen,

(f) between one and three, inclusive, of these events happen.

2 In a Poisson distribution the mean rate of occurrence of an event is 2.95 per 10 minutes. Find the probability that in a given 10 minutes:

(a) none of these events happen,

(b) exactly one of these events happens,

(c) exactly two of these events happen,

(d) exactly three of these events happen,

(e) more than one of these events happen,

(f) between one and three, inclusive, of these events happen.

3 In a Poisson distribution the mean rate of occurrence of an event is 0.84 per minute. Find the probability that in a given minute:

(a) none of these events happen,

(b) exactly one of these events happens,

(c) exactly two of these events happen,

(d) exactly three of these events happen,

(e) more than two of these events happen,

(f) between one and three, inclusive, of these events happen.

2

2.5 The sum of independent Poisson distributions

If cars, going north, pass a point on a motorway independently at random and cars, going south, on the same motorway pass the point independently at random then all cars on the motorway will pass the point independently at random. If the cars going north follow a Poisson distribution with mean 5 per minute and cars going south follow a Poisson distribution with mean 6 per minute then the total number of cars passing the point will follow a Poisson distribution with mean $5 + 6 = 11$ per minute. This is a particular example of the general result that:

> If X_1, X_2, X_3... follow independent Poisson distributions with means $\lambda_1, \lambda_2, \lambda_3$... respectively, then $X = X_1 + X_2 + X_3$... follows a Poisson distribution with mean
> $\lambda = \lambda_1 + \lambda_2 + \lambda_3$... .

In the example, above, of cars passing a point on a motorway, X_1 and X_2 are from different Poisson distributions. However, the result also applies if they are from the same distribution. For example X_1 could be the number of cars, going north, passing the point in a one minute interval and X_2 the number of cars, going north, passing the point in the next one minute interval. Both would be from a Poisson distribution with mean 5. $X_1 + X_2$ would be the number of cars, going north, passing the point in a 2-minute interval and would follow a Poisson distribution with mean $5 + 5 = 10$.

Worked example 2.4

The sales of a particular make of video recorder at two shops which are members of the same chain follow independent Poisson distributions with means 3 per day at the first shop and 4.5 per day at the second shop. Find the probability that on a given day:

(a) the first shop sells more than five,

(b) the second shop sells more than five,

(c) the total sales by the two shops is:

    (i) five or fewer,

    (ii) more than 10,

    (iii) between five and nine inclusive.

Solution

(a) $P(X_1 > 5) = 1 - P(X_1 \leqslant 5)$
$$= 1 - 0.9161$$
$$= 0.0839$$

(b) $P(X_2 > 5) = 1 - P(X_2 \leqslant 5)$
$$= 1 - 0.7029$$
$$= 0.297$$

(c) Total sales of the two shops is Poisson with mean $3 + 4.5 = 7.5$

    (i) $P(X \leqslant 5) = 0.2414$
$$= 0.241$$

    (ii) $P(X > 10) = 1 - P(X \leqslant 10)$
$$= 1 - 0.8622$$
$$= 0.1378$$
$$= 0.138$$

    (iii) $P(5 \leqslant X \leqslant 9) = P(X \leqslant 9) - P(X \leqslant 4)$
$$= 0.7764 - 0.1321$$
$$= 0.6443$$
$$= 0.644$$

Worked example 2.5

A 1-day course on statistics for teachers of A-level geography is first advertised 8 weeks before it is due to take place.

Throughout these 8 weeks, the number of places booked follows a Poisson distribution with mean 2 per week.

(a) Find the probability that, during the first week, two or fewer places are booked.

The organisers are hoping for at least 20 participants. They decide that, if at the end of the first 5 weeks less than 10 places have been booked, then they will cancel the guest speaker.

(b) Find the probability that the guest speaker will be cancelled.

(c) Find the probability of exactly nine places being booked during the first 5 weeks.

(d) Exactly nine places were booked during the first 5 weeks. Find the probability that sufficient places are booked in the remaining 3 weeks to give a total of 20 or more bookings during the 8 week period. [A]

Solution

(a) $P(X \le 2) = 0.6767$
$$= 0.677$$

(b) The number of bookings in 5 weeks will follow a Poisson distribution with mean $2 + 2 + 2 + 2 + 2 = 10$
$$P(X \le 9) = 0.4579$$
$$= 0.458 \rightarrow \text{course cancelled}$$

(c) $P(X = 9) = 0.4579 - 0.3328$
$$= 0.1251$$
$$= 0.125$$

(d) 20 or more in 8 weeks $\rightarrow$ 11 or more in last 3 weeks. Number of bookings in 3 weeks will follow a Poisson distribution with mean $3 \times 2 = 6$.
$$P(X \ge 11) = 1 - P(X \le 10)$$
$$= 1 - 0.9574$$
$$= 0.0426$$

EXERCISE 2D

1 The sales of cricket bats in a sports shop may be modelled by a Poisson distribution with mean 1.2 per day. Find the probability that:

(a) two or more bats are sold on a particular day,

(b) eight or more bats are sold in a 5-day period,

(c) exactly seven bats are sold in a 5-day period,

(d) between four and seven bats are sold in a 5-day period.

2 Calls arrive at a switchboard independently at random at an average rate of 1.4 per minute. Find the probability that:

(a) more than two calls will arrive in a particular minute,

(b) more than 10 calls will arrive in a 5-minute interval,

(c) between five and nine calls, inclusive, will arrive in a 5-minute interval,

(d) more than 20 calls will arrive in a 10-minute interval,

(e) 10 or fewer calls will arrive in a 10-minute interval.

3 The flaws in cloth produced on a loom may be modelled by a Poisson distribution with mean 0.4 per metre. Find the probability that there will be:

(a) two or fewer flaws in a metre of this cloth,

(b) more than three flaws in a 5-metre length of this cloth,

(c) between three and six flaws, inclusive, in a 10-metre length of this cloth,

(d) fewer than 10 flaws in a 30-metre roll of this cloth.

4 Two types of parasite were found on fish in a pond. They were distributed independently, at random with a mean of 0.8 per fish for the first type and 2.0 per fish for the second type. Find the probability that a fish will have:

(a) three or fewer parasites of the first type,

(b) more than one parasite of the second type,

(c) a total of three or fewer parasites,

(d) a total of exactly three parasites,

(e) a total of more than five parasites.

5 A garage has two branches. The sales of batteries may be modelled by a Poisson distribution with mean 2.4 per day at the first branch and by a Poisson distribution with mean 1.6 per day at the second branch. Find the probability that there will be:

(a) exactly three batteries sold at the first branch on a particular day,

(b) exactly three batteries sold at the second branch on a particular day,

(c) a total of five or more batteries sold at the two branches on a particular day,

(d) a total of between one and five, inclusive, batteries sold at the two branches on a particular day,

(e) 15 or fewer batteries sold at the first branch in a 5-day period,

(f) more than eight batteries sold at the second branch in a 5-day period.

2.6 Using the Poisson distribution as a model

The Poisson distribution occurs when events occur independently, at random, at a constant average rate. A common example is cars passing a point on a motorway. However, if the motorway was busy, cars would obstruct each other and so would not pass at random. Hence the Poisson distribution would not be a suitable model. Also if we observed over a 24-hour period, the mean number of cars per minute would not be constant. In the middle of the night the mean would be less than in the middle of the day. Again the Poisson distribution would not provide a suitable model. It probably would provide a suitable model if we counted the number of cars per minute on a free-flowing motorway over a relatively short period of time.

Telephone calls arriving at a switchboard are also often modelled by a Poisson distribution. However, if there were a queueing system or the switchboard was frequently engaged the calls would not be arriving independently at random. Also over a 24-hour period the mean would change and so the Poisson distribution would not be a suitable model. The Poisson distribution will often provide an adequate model for people joining queues in a supermarket or at a train station. However, if a family of four are shopping together they will probably all join the queue at the same time and so events will not be independent. Again if the queue is long people may be deterred from entering the supermarket and so once again the events would not be independent.

Despite all these qualifications the Poisson distribution provides a useful model in many practical situations. As with any probability distribution, we can never prove that events follow a Poisson distribution exactly, but we can recognise circumstances where it is likely to provide an adequate model.

EXERCISE 2E

State whether or not the Poisson distribution is likely to provide a suitable model for the random variable, X, in the following examples. Give a reason where you believe the Poisson distribution would not provide a suitable model.

The random variable, X, represents number of:

(a) lorries per minute passing a point on a quiet motorway over a short period of time,

(b) lorries per minute passing a point on a very busy motorway,

(c) cars per minute passing a point close to traffic lights on a city centre road,

(d) components which do not meet specification in a sample of 20 from a production line,

(e) dandelions in a square metre of lawn in a small garden,

(f) boxes of expensive chocolates sold per day at a small shop,

(g) passengers per minute arriving at a bus stop, over a short period of time,

(h) passengers per minute arriving at a bus stop over a 24-hour period,

(i) breakdowns of a power supply per year,

(j) accidents per year in a large factory,

(k) people injured per year in accidents at a large factory.

Worked example 2.6

Travellers arrive at a railway station, to catch a train, either alone or in family groups. On an August Saturday afternoon, the number, X, of travellers who arrive alone during a 1-minute interval may be modelled by a Poisson distribution with mean 7.5.

(a) Find the probability of six or fewer passengers arriving alone during a particular minute.

The number, Y, of family groups who arrive during a one-minute interval may be modelled by a Poisson distribution with mean 2.0.

(b) Find the probability that three or more family groups arrive during a particular minute.

It is usual for one person to buy all the tickets for a family group. Thus the number of people, Z, wishing to buy tickets during a one-minute interval may be modelled by $X + Y$.

(c) Find the probability that more than 18 people wish to buy tickets during a particular minute.

If four booking clerks are available, they can usually sell tickets to up to 18 people during a minute.

(d) State, giving a reason in **each** case, whether:

    (i) more than four booking clerks should be available on an August Saturday afternoon.

    (ii) the Poisson distribution is likely to provide an adequate model for the total number of travellers (whether or not in family groups) arriving at the station during a 1-minute interval.

(iii) the Poisson distribution is likely to provide an adequate model for the number of passengers, travelling alone, leaving the station, having got off a train, during a 1-minute interval.

(e) Give **one** reason why the model $Z = X + Y$, used in **(c)**, may not be exact.

Solution

(a) $P(X \leqslant 6) = 0.3782$
$\qquad = 0.378$

(b) $P(X \geqslant 3) = 1 - P(\leqslant 2)$
$\qquad\qquad = 1 - 0.6767 = 0.3233$
$\qquad\qquad = 0.323$

(c) $X + Y \rightarrow$ Poisson mean $7.5 + 2 = 9.5$
$\qquad P(X > 18) = 1 - P(X \leqslant 18)$
$\qquad\qquad\qquad = 1 - 0.9957 = 0.0043$

> Only two significant figures in the answer can be obtained from the tables. In these circumstances a two significant figure answer will be accepted in an examination.

(d) **(i)** No, the probability of more than 18 people wishing to buy tickets in a minute has been shown in **(c)** to be very small. Hence four booking clerks should be adequate.

(ii) No, the people arriving in family groups will not be arriving independently. Hence Poisson unlikely to be an adequate model.

(iii) No, the average rate will not be constant. Immediately after a train arrives there will be a high average rate, between train arrivals there will be a low average rate.

(e) Some people or groups may have bought tickets in advance/some family groups may buy tickets individually.

2.7 Variance of a Poisson distribution

The Poisson distribution has the interesting property that the variance is equal to the mean. That is, a Poisson distribution, with mean 9, will have a variance of 9. You will probably be more interested in the standard deviation which will be $\sqrt{9} = 3$.

> You will remember from MS01 that the variance is used by theoretical statisticians but is a poor measure of spread.

> A Poisson distribution with mean λ has a variance of λ (and a standard deviation of $\sqrt{\lambda}$).

The number of items of post delivered to a particular address, daily, follows a Poisson distribution with mean 9. On 10 days the number of items delivered was

$\qquad$ 10 $\quad$ 8 $\quad$ 13 $\quad$ 9 $\quad$ 4 $\quad$ 9 $\quad$ 12 $\quad$ 15 $\quad$ 11 $\quad$ 8

> mean $\bar{x} = 9.9$
> standard deviation $s = 3.07$
> variance $s^2 = 9.43$

This is a sample from a Poisson distribution with mean 9 and so the mean of the sample will almost certainly not be exactly 9 nor will the variance be exactly 9. However for a large sample we would expect the mean and the variance both to be very close to 9.

This result will be used in later modules but for now its only application is to provide you with an extra piece of information when deciding whether or not the Poisson distribution may provide an adequate model. If you expected the number of items of post delivered per day to follow a Poisson distribution, then the fact that the mean (9.9) and variance (9.43) of the sample were close together would support this. Suppose the number of letters delivered on a sample of ten days had been:

   12   6   15   7   2   7   14   17   13   6

with a mean of 9.9 and a variance of 24.1. You would have to say that it was very unlikely that the Poisson distribution would provide an adequate model as the mean and variance are so far apart.

MIXED EXERCISE

1 A small shop stocks expensive boxes of chocolates whose sales may be modelled by a Poisson distribution with mean 1.8 per day. Find the probability that on a particular day the shop will sell

   (a) no boxes,

   (b) three or more boxes of these chocolates.   [A]

2 The number of births announced in the personal column of a local weekly newspaper may be modelled by a Poisson distribution with mean 2.4.

   Find the probability that, in a particular week:

   (a) three or fewer births will be announced,

   (b) exactly four births will be announced.   [A]

3 The number of customers entering a certain branch of a bank on a Monday lunchtime may be modelled by a Poisson distribution with mean 2.4 per minute.

   Find the probability that, during a particular minute, four or more customers enter the branch.   [A]

4 A shop sells a particular make of video recorder.

   (a) Assuming that the weekly demand for the video recorder is a Poisson variable with mean 3, find the probability that the shop sells:
      (i)   at least three in a week,
      (ii) at most seven in a week,
      (iii) more than 20 in a month (4 weeks).

Stocks are replenished only at the beginning of each month.

(b) Find the minimum number that should be in stock at the beginning of a month so that the shop can be at least 95% sure of being able to meet the demand during the month.

5 Incoming telephone calls to a school arrive at random times. The average rate will vary according to the day of the week. On Monday mornings in term time there is a constant average rate of four per hour. What is the probability of receiving:

(a) six or more calls in a particular hour,

(b) three or fewer calls in a particular period of two hours?

During term time on Friday afternoons the average rate is also constant and it is observed that the probability of no calls being received during a particular hour is 0.202. What is the average rate of calls on Friday afternoons? [A]

6 State giving a reason whether or not the Poisson distribution is likely to provide an adequate model for the following distributions.

(a) A transport cafe is open 24 hours a day. The number of customers arriving in each 5-minute period of a particular day is counted.

(b) Following a cup semi-final victory, a football club ticket office receives a large number of telephone enquiries about tickets for the final, resulting in the switchboard frequently being engaged. The number of calls received during each 5-minute period of the first morning after the victory is recorded.

(c) A machine produces a very large number of components of which a small proportion are defective. At regular intervals samples of 150 components are taken and the number of defectives counted.

7 A car-hire firm finds that the daily demand for its cars follows a Poisson distribution with mean 3.6.

(a) What is the probability that on a particular day the demand will be:
  (i) two or fewer,
  (ii) between three and seven (inclusive),
  (iii) zero?

(b) What is the probability that 10 consecutive days will include two or more on which the demand is zero?

(c) Suggest reasons why daily demand for car hire may not follow a Poisson distribution. [A]

8 The number of letters received by a household on a weekday follows a Poisson distribution with mean 2.8.

(a) What is the probability that on a particular weekday the household receives three or more letters.

(b) Explain briefly why a Poisson distribution is unlikely to provide an adequate model for the number of letters received on a weekday throughout the year. [A]

9 The number of telephone calls to a university admissions office is monitored.

(a) During working hours in January the number of calls received follows a Poisson distribution with mean 1.8 per 15-minute interval. During a particular 15-minute interval:

(i) what is the probability that two or fewer calls are received,

(ii) what number of calls is exceeded with probability just greater than 0.01?

(b) On any particular working day the number of attempts to telephone the office is distributed at random at a constant average rate. Usually an adequate number of staff are available to answer the telephone. However, for a short period in August, immediately after the publication of A-level results, the number of calls increases and the telephones are frequently engaged.

State, giving a reason, whether the Poisson distribution is likely to provide an adequate model for each of the following distributions:

(i) the number of calls received in each minute during working hours of a day in June,

(ii) the number of calls received in each minute during working hours of a day immediately after the publication of A-level results,

(iii) the number of calls received on each working day throughout the year. [A]

10 Bronwen runs a post office in a large village. The number of registered letters posted at this office may be modelled by a Poisson distribution with mean 1.4 per day.

(a) Find the probability that at this post office:

(i) two or fewer registered letters are posted on a particular day,

(ii) a total of four or more registered letters are posted on two consecutive days.

(b) The village also contains a post office run by Gopal. Here the number of registered letters posted may be modelled by a Poisson distribution with mean 2.4 per day. Find the probability that, on a particular day, the number of registered letters posted at Gopal's post office is less than 4.

(c) The numbers of registered letters posted at the two post offices are independent. Find the probability that, on a particular day, the total number of registered letters posted at the two post offices is more than six.

(d) Give one reason why the Poisson distribution might not provide a suitable model for the number of registered letters posted daily at a post office. [A]

11 The number of cars, travelling from East to West, passing a point on a motorway, may be modelled by a Poisson distribution with a mean of 1.2 per five-second interval.

(a) Find the probability that, during a particular five-second interval, the number of cars, travelling from East to West, which pass the point is
  (i) zero,
  (ii) exactly two.

The number of cars, travelling from West to East, passing the same point on the motorway, may be modelled by a Poisson distribution with a mean of 3.8 per five-second interval.

(b) Find the probability that, during a particular five-second interval, more than eight cars, travelling from West to East, pass the point.

(c) Find the probability that, during a particular five-second interval, the total number of cars passing the point is less than eight. (You may assume that the number of cars travelling from East to West is independent of the number of cars travelling from West to East.)

(d) Explain why a Poisson distribution may not provide an adequate model for the total number of car passengers passing the point in a five-second interval. [A]

12 The number of vehicles arriving at a toll bridge during a 5-minute period can be modelled by a Poisson distribution with mean 3.6.

(a) State the value for the standard deviation of the number of vehicles arriving at the toll during a 5-minute period.

(b) Find:
  (i) the probability that at least three vehicles arrive in a 5-minute period,
  (ii) the probability that at least three vehicles arrive in each of three successive 5-minute periods.

(c) Show that the probability that no vehicles arrive in a 10-minute period is 0.0007, correct to four decimal places.

Key point summary

1 The Poisson distribution is the distribution of events   *p27*
which occur independently, at random, at a constant
average rate.

2 For a constant average rate λ, the probability of r *p31*

events occurring, $P(R = r) = e^{-\lambda}\dfrac{\lambda^r}{r!}$

3 Dependent on the value of λ, Poisson probabilities   *p28*
may be found from tables.

4 If $X_1, X_2, X_3 \ldots$ follow independent Poisson   *p33*
distributions with means $\lambda_1, \lambda_2, \lambda_3 \ldots$, respectively,
then $X = X_1 + X_2 + X_3\ldots$ follows a Poisson distribution
with mean $\lambda = \lambda_1 + \lambda_2 + \lambda_3\ldots$.

5 A Poisson distribution with mean λ has a variance   *p39*
of λ.

Test yourself	What to review
1 A Poisson distribution has a mean of 1.2 events per minute. Find, from tables, the probability that in a particular minute	*Section 2.3*
(a) three or fewer events occur,	
(b) exactly three events occur,	
(c) less than three events occur,	
(d) more than three events occur.	
2 What is the largest number of events which could occur in a given minute for the Poisson distribution in question 1?	*Section 2.1*
3 A Poisson distribution has mean 1.12 events per minute. Calculate the probability that in a particular minute	*Section 2.4*
(a) exactly one event will occur,	
(b) less than two events will occur,	
(c) more than two events will occur.	
4 The number of calls received at a switchboard may be modelled by a Poisson distribution with mean 12 per hour. Find the probability that more than one call will be received in a particular 5-minute interval.	*Section 2.3*
5 State two conditions which must be fulfilled if the number of bicycles crossing a bridge per minute is to follow a Poisson distribution.	*Sections 2.1 and 2.6*

Test yourself (*continued*)	**What to review**

6 Explain why the Poisson distribution is unlikely to form an adequate model for the number of bicycles crossing a bridge   *Sections 2.1 and 2.6*

  (a) over a 24-hour period,

  (b) in the rush hour when a large number of bicycles are attempting to cross.

---

7 The number of newspapers sold by a newsagent in eight successive hours on a weekday was   *Sections 2.6 and 2.7*

       84   92   22   12   13   9   8   104.

  (a) Calculate the mean and variance of the data.

  (b) Give a reason based on your calculations why it is unlikely that the Poisson distribution will provide an adequate model for the data.

  (c) Give a reason, not based on your calculations, why it is unlikely that a Poisson distribution will provide an adequate model for the hourly sales of newspapers throughout a day.

Test yourself   **ANSWERS**

1 **(a)** 0.966;   **(b)** 0.0867;   **(c)** 0.8795;   **(d)** 0.0338.

2 In theory there is no upper limit.

3 **(a)** 0.365;   **(b)** 0.692;   **(c)** 0.104.

4 0.264.

5 Constant average rate; cross independently.

6 **(a)** The average rate will not be constant over a 24-hour period.

  (b) Cycles will obstruct each other so the crossings will not be independent.

7 **(a)** Mean 43.0, variance 1783.7.

  (b) Large difference between mean and variance.

  (c) Mean unlikely to be constant. More papers sold as people travel to and from work in morning and evening than in the middle of the day.

CHAPTER 3

Continuous probability distributions

Learning objectives

After studying this chapter, you should be able to:

- understand what is meant by a continuous random variable
- understand the relationship between the probability density function, f(x), and the cumulative distribution function, F(x), of a continuous random variable, X
- find the probability of an observation lying in a specified interval
- find the median, quartiles and percentiles of a continuous random variable
- find the mean, standard deviation and variance of a continuous random variable
- find the mean, standard deviation and variance of a simple function of a continuous random variable
- use the rectangular distribution.

3.1 Introduction

In S1, you met the most important continuous probability distribution: the normal distribution.

You have already seen in Chapter 1, that for a discrete random variable X, it is possible to allocate probabilities to each discrete value, x, that X can take. However, this is **not** the case for a continuous random variable.

For a continuous random variable X, we allocate probabilities to each of the **range of values** that the variable can take.

This is usually achieved by defining a simple function f(x), called the **probability density function**.

3.2 Continuous random variables

Probability density function (p.d.f.)

The probability density function, f(x), which allocates probabilities to each of the range of values that the continuous random variable can take, is such that:

The graph of $y = $ f(x) always lies on or above the x-axis.

$\qquad$ f(x) $\geqslant 0$, for all values of x.

$$\int_{\text{all } x} f(x)\, dx = 1$$

> The probability that the random variable X takes a value in the range $a < x < b$ or $a \leqslant x \leqslant b$ is given by:
>
> $$P(a < X < b) = P(a \leqslant X \leqslant b) = \int_{x = a}^{x = b} f(x)\, dx$$

In the examination, f(x) will be a simple polynomial.

Worked example 3.1

A continuous random variable, X, has the probability density function defined by:

$$f(x) = \begin{cases} k(1 - x^2) & -1 \leqslant x \leqslant 1 \\ 0 & \text{otherwise,} \end{cases}$$

where k is a constant.

(a) Sketch f(x).　　　　**(b)** Show that $k = \dfrac{3}{4}$.

(c) Calculate:

(i) P($X = 0.5$),　　　　**(ii)** P($0.2 < X < 0.5$).

Solution

(a)

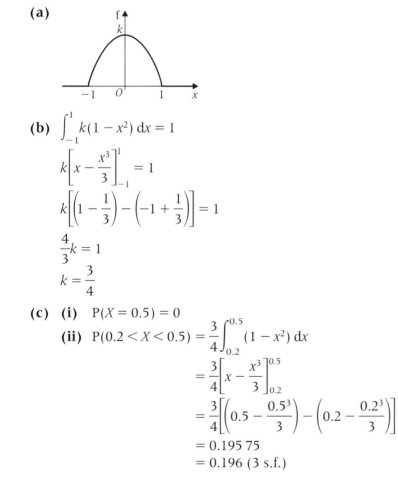

(b) $\displaystyle\int_{-1}^{1} k(1 - x^2)\, dx = 1$

$$k\left[x - \frac{x^3}{3} \right]_{-1}^{1} = 1$$

$$k\left[\left(1 - \frac{1}{3}\right) - \left(-1 + \frac{1}{3}\right) \right] = 1$$

$$\frac{4}{3}k = 1$$

$$k = \frac{3}{4}$$

The total area enclosed by the graph of $y = f(x)$ and the x-axis, over all the values for which f(x) is defined, is always equal to one, i.e. $\int_{\text{all } x}^{1} f(x)\, dx = 1$.

(c) **(i)** P($X = 0.5$) = 0

$P(X = c) = 0$ for all values of a constant c.

(ii) $\displaystyle P(0.2 < X < 0.5) = \frac{3}{4}\int_{0.2}^{0.5} (1 - x^2)\, dx$

$$= \frac{3}{4}\left[x - \frac{x^3}{3} \right]_{0.2}^{0.5}$$

$$= \frac{3}{4}\left[\left(0.5 - \frac{0.5^3}{3} \right) - \left(0.2 - \frac{0.2^3}{3} \right) \right]$$

$$= 0.195\,75$$

$$= 0.196 \text{ (3 s.f.)}$$

Using:
$P(a < X < b) = \int_a^b f(x)\, dx$, where
$a = 0.2$ and $b = 0.5$.

3

Worked example 3.2

The continuous random variable X has the following probability density function:

$$f(x) = \begin{cases} \dfrac{1}{8}(x + 2) & -2 \leqslant x \leqslant 2 \\ 0 & \text{otherwise.} \end{cases}$$

(a) Sketch the graph of f.

(b) Write down the value of $P(X = 1)$.

(c) Calculate $P(-1 < x < 1)$.

Solution

(a)

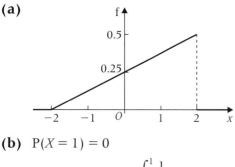

(b) $P(X = 1) = 0$

$P(X = c) = 0$ for all constants c.

(c) $P(-1 < X < 1) = \displaystyle\int_{-1}^{1} \dfrac{1}{8}(x + 2)\, \mathrm{d}x$

$$= \left[\dfrac{x^2}{16} + \dfrac{x}{4} \right]_{-1}^{1}$$

$$= 0.5$$

Alternative solution

$$f(-1) = \dfrac{1}{8}$$

$$f(1) = \dfrac{3}{8}$$

You can now find the area of the trapezium.

$$\text{Area} = \dfrac{1}{2}[f(1) + f(-1)] \times 2$$

$$= \dfrac{1}{2} \times \dfrac{4}{8} \times 2$$

$$= 0.5$$

When $f(x)$ is either fully or partially defined by a straight line, it is often easier to calculate probabilities by finding areas of rectangles, triangles and trapezia, as shown in the alternative solution.

Worked example 3.3

A continuous random variable, X, has a probability density function defined by

$$f(x) = \begin{cases} ax(4-x) & 0 \leqslant x \leqslant 4 \\ 0 & \text{otherwise,} \end{cases}$$

where a is a constant.

(a) Show that $a = \dfrac{3}{32}$.

(b) Sketch the graph of f.

(c) Calculate $P(1 \leqslant X \leqslant 3)$.

Solution

(a) $\displaystyle\int_0^4 f(x)\, dx = 1$

$\displaystyle\int_0^4 ax(4-x)\, dx = 1$

$a\left[2x^2 - \dfrac{x^3}{3} \right]_0^4 = 1$

$a\left[\left(32 - \dfrac{64}{3} \right) - (0) \right] = 1$

$\dfrac{32}{3}a = 1$

$a = \dfrac{3}{32}$

> Using:
>
> $$\int_{\text{all } x} f(x)\, dx = 1.$$

(b)

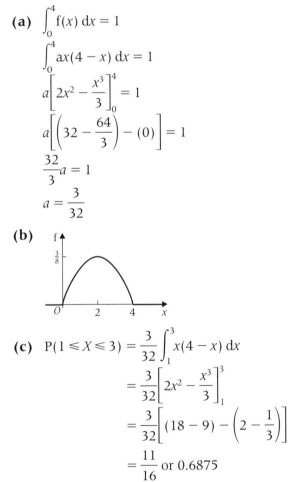

(c) $P(1 \leqslant X \leqslant 3) = \dfrac{3}{32} \displaystyle\int_1^3 x(4-x)\, dx$

$= \dfrac{3}{32}\left[2x^2 - \dfrac{x^3}{3} \right]_1^3$

$= \dfrac{3}{32}\left[(18-9) - \left(2 - \dfrac{1}{3} \right) \right]$

$= \dfrac{11}{16}$ or 0.6875

EXERCISE 3A

1 A continuous random variable, X, has the following probability density function:

$$f(x) = \begin{cases} kx & 1 < x < 4 \\ 0 & \text{otherwise,} \end{cases}$$

where k is a constant.

   (a) Find the value of k.

   (b) Sketch the graph of f.

   (c) Write down the value of $P(X = 1.5)$.

   (d) Calculate $P(2 < X < 3)$.

2 A continuous random variable, Y, has a probability density function defined by

$$f(y) = \begin{cases} ay^2 & 0 < y < 4 \\ 0 & \text{otherwise,} \end{cases}$$

where a is a constant.

   (a) Show that $a = \dfrac{3}{64}$.

   (b) Sketch the graph of f.

   (c) Write down the value of $P(Y = 2)$.

   (d) Calculate $P(1 \leqslant Y < 3)$.

3 A continuous random variable, R, has the following probability density function:

$$f(r) = \begin{cases} \frac{2}{3}r & 0 < r \leqslant 1 \\ \frac{2}{27}(r - 4)^2 & 1 < r \leqslant 4 \\ 0 & \text{otherwise.} \end{cases}$$

   (a) Sketch the graph of f.

   (b) Calculate $P(0.5 < R < 3)$.

4 A major road junction delays motorists on their way to work. The time, T minutes, by which their journeys are delayed is a continuous random variable with a probability density function defined by

$$f(t) = \begin{cases} k(1 + 4t) & 0 < t \leqslant 30 \\ 0 & \text{otherwise,} \end{cases}$$

where k is a constant.

   (a) Find the value of k.

   (b) Find the probability that a motorist selected at random, from those using this road junction, will experience a delay of:

      (i) exactly 10 minutes,

      (ii) between 20 and 30 minutes.

5 A random variable X has the following probability density function:

$$f(x) = \begin{cases} \dfrac{1}{8} & -2 \leqslant x < 0 \\ \dfrac{1}{4} & 0 \leqslant x < 3 \\ 0 & \text{otherwise.} \end{cases}$$

   (a) Sketch the graph of f.

   (b) Find $P(-1 < X < 2)$.

3.3 The cumulative distribution function

The cumulative distribution function, $F(x)$, for a continuous random variable, X, having probability density function $f(x)$ is such that:

> For any value x of X, in the range of values for which $f(x)$ is defined,
>
> $$F(x) = P(X \leq x) = \int_{-\infty}^{x} f(x)\, dx$$
>
> and
>
> $$\frac{d}{dx}F(x) = f(x)$$

If $f(x)$ is defined on the range of values $a \leq x \leq b$, then

$$F(x) = P(X \leq x) = \int_{a}^{x} f(x)\, dx.$$

3

Worked example 3.4

The random variable X has the cumulative distribution function defined by

$$F(x) = \begin{cases} 0 & x < -2 \\ \dfrac{1}{8}(x + 2) & -2 \leq x < 0 \\ \dfrac{1}{4}(x + 1) & 0 \leq x < 3 \\ 1 & x \geq 3. \end{cases}$$

Sketch the graph of F.

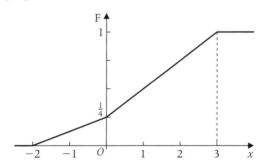

In general, F is a continuous function where $0 \leq F(x) \leq 1$.

When you draw the graph of $F(x)$, a good point to note is that as the value x, of the random variable X, increases, the graph of $F(x)$ **never** decreases.

Worked example 3.5

The continuous random variable X has the following cumulative distribution function:

$$F(x) = \begin{cases} 0 & x \leq 0 \\ \dfrac{x^3}{64} & 0 < x \leq 4 \\ 1 & x > 4. \end{cases}$$

(a) Find $P(X \leq 3)$.

(b) Sketch the graph of f.

Solution

(a) $P(X \leq 3) = F(3)$

$$= \frac{3^3}{64}$$

$$= \frac{27}{64}$$

(b) Using $f(x) = \dfrac{d}{dx}F(x)$, you can find the probability density function to be defined as

$$f(x) = \begin{cases} \dfrac{3x^2}{64} & 0 \leq x \leq 4 \\ 0 & \text{otherwise.} \end{cases}$$

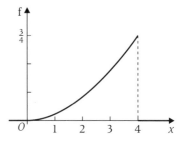

Worked example 3.6

A continuous random variable X has probability density function $f(x)$ defined by

$$f(x) = \begin{cases} \dfrac{1}{8}(x + 2) & -2 \leq x \leq 2 \\ 0 & \text{otherwise.} \end{cases}$$

(a) Sketch the graph of f.

(b) Find the cumulative distribution function $F(x)$.

(c) Sketch the graph of F.

(d) Find $P(X \leq 1)$.

Solution

(a)

(b) For the range of values $-2 \leq x \leq 2$

$$F(x) = \frac{1}{8}\int_{-2}^{x} (x + 2)\, dx$$

$$= \frac{1}{16}(x + 2)^2$$

> **Using the calculus result**
> $$\int (a + x)^n\, dx = \frac{(a + x)^{n+1}}{n+1}.$$

Hence, $F(x)$ can now be defined as follows:

$$F(x) = \begin{cases} 0 & x < -2 \\ \dfrac{1}{16}(x + 2)^2 & -2 \leq x \leq 2 \\ 1 & x > 2 \end{cases}$$

(c)

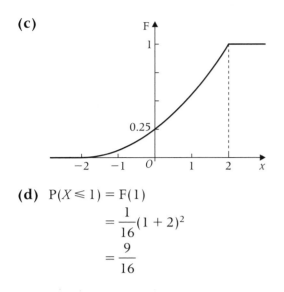

3

(d) $P(X \leqslant 1) = F(1)$

$$= \frac{1}{16}(1 + 2)^2$$

$$= \frac{9}{16}$$

Worked example 3.7

The continuous random variable Y has a probability density function defined by

$$f(y) = \begin{cases} \dfrac{3}{64}y^2 & 0 < y < 4 \\ 0 & \text{otherwise.} \end{cases}$$

(a) Find the cumulative distribution function $F(y)$.

(b) Sketch the graph of F.

(c) Find $P(Y \geqslant 2)$.

Solution

(a) For the range of values $0 < y < 4$

$$F(y) = \int_0^y \frac{3}{64}y^2 \, dy$$

$$= \frac{1}{64}y^3$$

Hence, $F(y)$ can now be defined as follows:

$$F(y) = \begin{cases} 0 & y \leqslant 0 \\ \dfrac{1}{64}y^3 & 0 < y < 4 \\ 1 & y \geqslant 4. \end{cases}$$

(b)

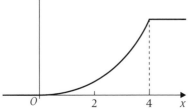

(c) $P(Y \geqslant 2) = 1 - P(Y < 2)$
$$= 1 - F(2)$$
$$= 1 - \frac{8}{64}$$
$$= \frac{7}{8}$$

Worked example 3.8

The continuous random variable R has the probability density function defined by:

$$f(r) = \begin{cases} \dfrac{1}{9}r & 0 \leqslant r < 3 \\ \dfrac{1}{9}(6 - r) & 3 \leqslant r \leqslant 6 \\ 0 & \text{otherwise.} \end{cases}$$

(a) Sketch the graph of f.

(b) Find the cumulative distribution function $F(r)$.

(c) Calculate:

    (i) $P(R \leqslant 3)$;   **(ii)** $P(2 \leqslant R \leqslant 5)$.

(d) Sketch the graph of F.

Solution

(a)

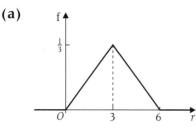

(b) Since $f(r)$ is given in two parts, one for $0 \leqslant r < 3$ and one for $3 \leqslant r \leqslant 6$, we must find F by using two separate calculations.

For $0 \leqslant r < 3$,

$$F(r) = \int_0^r \frac{1}{9}r \, dr = \frac{1}{18}r^2$$

For $3 \leqslant r \leqslant 6$,

$$P(R \leqslant r) = F(3) + \int_3^r \frac{1}{9}(6 - r) \, dr$$

$$= 0.5 + \left[-\frac{1}{18}(6 - r)^2 \right]_3^r$$

$$= 0.5 + \left[0.5 - \frac{1}{18}(6 - r)^2 \right]$$

$$= 1 - \frac{1}{18}(6 - r)^2$$

$$\therefore \quad F(r) = \begin{cases} 0 & r < 0 \\ \dfrac{1}{18}r^2 & 0 \leqslant r < 3 \\ 1 - \dfrac{1}{18}(6-r)^2 & 3 \leqslant r \leqslant 6 \\ 1 & r > 6 \end{cases}$$

(c) (i) $P(R \leqslant 3) = F(3) = \dfrac{1}{18} \times 3^2 = 0.5$

(ii) $P(2 \leqslant R \leqslant 5) = F(5) - F(2)$

$$= \frac{17}{18} - \frac{4}{18}$$

$$= \frac{13}{18} \text{ or } 0.722$$

(d)

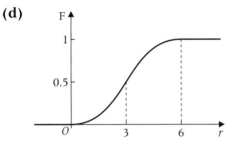

EXERCISE 3B

1 The continuous random variable X has a cumulative distribution function defined by

$$F(x) = \begin{cases} 0 & x < 0 \\ \dfrac{x^3}{125} & 0 \leqslant x \leqslant 5 \\ 1 & x > 5. \end{cases}$$

(a) Write down the value of $P(X = 4)$.

(b) Find $P(X \leqslant 3)$.

(c) Find $f(x)$.

(d) Sketch the graph of $f(x)$.

2 The continuous random variable Y has the following probability density function:

$$f(y) = \begin{cases} \dfrac{1}{8} & 0 < y < 2 \\ \dfrac{1}{96}(14 - y) & 2 \leqslant y < 14 \\ 0 & \text{otherwise.} \end{cases}$$

(a) Sketch the graph of f.

(b) Find $F(y)$.

(c) Sketch the graph of F.

(d) Calculate the value of $P(Y \leqslant 8)$.

3 The continuous random variable T has probability density function defined by

$$f(t) = \begin{cases} kt(6-t) & 3 \leqslant t \leqslant 6 \\ 0 & \text{otherwise.} \end{cases}$$

(a) Show that $k = \dfrac{1}{18}$.

(b) Sketch the graph of f.

(c) Find $F(t)$.

(d) Sketch the graph of F.

(e) Calculate $P(T \leqslant 4)$.

4 The continuous random variable R has the following probability density function:

$$f(r) = \begin{cases} \dfrac{2}{3}r & 0 \leqslant r < 1 \\ \dfrac{2}{27}(r-4)^2 & 1 \leqslant r \leqslant 4 \\ 0 & \text{otherwise.} \end{cases}$$

(a) Find the cumulative distribution function $F(r)$.

(b) Calculate $P(R \geqslant 3)$.

(c) Find $P(0.5 < R < 1.5)$.

5 A random variable X has cumulative distribution function defined by:

$$F(x) = \begin{cases} 0 & x \leqslant -2 \\ \dfrac{1}{16}(x+2) & -2 < x \leqslant 0 \\ \dfrac{1}{16}(3x+2) & 0 < x \leqslant 2 \\ \dfrac{1}{8}(x+2) & 2 < x \leqslant 6 \\ 1 & x > 6 \end{cases}$$

(a) Calculate $P(X \leqslant 4)$.

(b) Sketch the graph of f.

3.4 The median, quartiles and percentiles

The median, m, of a continuous random variable X having probability density function f(x) is such that:

$$F(m) = P(X \leqslant m) = \int_{-\infty}^{m} f(x)\, dx = 0.5$$

If f(x) is defined on the range of values $a \leqslant x \leqslant b$ then
$F(m) = P(X \leqslant m) = 0.5$
$\Rightarrow \int_{a}^{m} f(x)\, dx = 0.5$

Worked example 3.9

The continuous random variable X has the following cumulative distribution function, $F(x)$:

$$F(x) = \begin{cases} 0 & x < 0 \\ \dfrac{1}{8}x^3 & 0 \leqslant x \leqslant 2 \\ 1 & x > 2 \end{cases}$$

Determine the value of the median.

3

Solution

$$F(m) = \frac{1}{8}m^3 = 0.5$$
$$m^3 = 4$$
$$m = \sqrt[3]{4}$$
$$= 1.587$$

Worked example 3.10

The continuous random variable X has probability density function defined by:

$$f(x) = \begin{cases} \dfrac{1}{12} & -8 \leqslant x < 0 \\ \dfrac{1}{48}(3x^2 + 4) & 0 \leqslant x < 2 \\ 0 & \text{otherwise.} \end{cases}$$

(a) Sketch the graph of f.

(b) Find the cumulative distribution function $F(x)$ for $-8 \leqslant x < 0$.

(c) Calculate $P(X \leqslant 0)$

(d) Determine the value of the median, m.

Solution

(a)

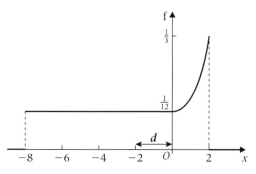

(b) For the range of values $-8 \leqslant x < 0$

$$F(x) = \int_{-8}^{x} \frac{1}{12} \, dx$$

$$= \left[\frac{1}{12} x \right]_{-8}^{x}$$

$$= \frac{1}{12} x + \frac{2}{3}$$

$$= \frac{1}{12} (x + 8)$$

(c) $P(X \leqslant 0) = F(0) = \dfrac{2}{3}$

(d) $F(m) = 0.5$

$\therefore$ m lies in the range $-8 \leqslant x < 0$

$$F(m) = \frac{1}{12} (m + 8) = 0.5$$

$$m + 8 = 6$$

$$m = -2$$

> The median could have been found very easily by using the graph of f(x).
>
> Since $F(0) = \dfrac{2}{3}$, it is obvious that the median lies in the range $-8 \leqslant x < 0$. You can now use the area of a rectangle to find the value of m.
>
> $$\frac{1}{12} \times d = \frac{2}{3} - \frac{1}{2}$$
>
> $$\Rightarrow \quad d = 12 \times \frac{1}{6} = 2$$
>
> $\therefore$ the value of the median, $m = -2$.

> The lower quartile q_1 and upper quartile q_3 of a continuous random variable X having probability density function $f(x)$ is such that
>
> $$F(q_1) = P(X \leqslant q_1) = \int_{-\infty}^{q_1} f(x) \, dx = 0.25$$
>
> and
>
> $$F(q_3) = P(X \leqslant q_3) = \int_{-\infty}^{q_3} f(x) \, dx = 0.75.$$

> If f(x) is defined on the range of values $a \leqslant x \leqslant b$ then
>
> $$F(q_1) = P(X \leqslant q_1) = 0.25$$
>
> $$\Rightarrow \int_{a}^{q_1} f(x)\, dx = 0.25$$
>
> $$F(q_3) = P(X \leqslant q_3) = 0.75$$
>
> $$\Rightarrow \int_{a}^{q_3} f(x)\, dx = 0.75$$

Worked example 3.11

A continuous random variable X has the cumulative distribution function defined by

$$F(x) = \begin{cases} 0 & x < 0 \\ \dfrac{1}{27} x^3 & 0 \leqslant x \leqslant 3 \\ 1 & x > 3 \end{cases}$$

Find the lower quartile, q_1, and the upper quartile, q_3, of X.

Solution

Lower quartile:

$$F(q_1) = \frac{1}{27} q_1^3 = 0.25$$

$$\Rightarrow \quad q_1^3 = 6.75$$

$$q_1 = \sqrt[3]{6.75} = 1.89$$

Upper quartile:

$$F(q_3) = \frac{1}{27} q_3^3 = 0.75$$

$$\Rightarrow \quad q_3^3 = 20.25$$

$$q_3 = \sqrt[3]{20.25} = 2.73$$

Worked example 3.12

The continuous random variable R has the following probability density function:

$$f(r) = \begin{cases} \dfrac{1}{16}(r+2) & 0 \leqslant r \leqslant 4 \\ 0 & \text{otherwise.} \end{cases}$$

Determine the values of the lower and upper quartiles of R.

Solution

For the range of values $0 \leqslant r \leqslant 4$

$$F(r) = \int_0^r \frac{1}{16}(r+2) = \frac{1}{32}r(r+4).$$

Since q_1 and q_3 have to lie in the range $0 \leqslant r \leqslant 4$,

$$F(q_1) = \frac{1}{32}q_1(q_1 + 4) = 0.25 \qquad\qquad F(q_3) = \frac{1}{32}q_3(q_3 + 4) = 0.75$$

$$\Rightarrow \quad q_1(q_1 + 4) = 8 \qquad\qquad\qquad \Rightarrow \quad q_3(q_3 + 4) = 24$$

$$q_1^2 + 4q_1 - 8 = 0 \qquad\qquad\qquad\quad q_3^2 + 4q_3 - 24 = 0$$

$$q_1 = \frac{-4 \pm \sqrt{16 - (-32)}}{2} \qquad\qquad q_3 = \frac{-4 \pm \sqrt{16 - (-96)}}{2}$$

$$= \frac{-4 \pm 4\sqrt{3}}{2} \qquad\qquad\qquad\quad = \frac{-4 \pm 4\sqrt{7}}{2}$$

$$= -2 \pm 2\sqrt{3} \qquad\qquad\qquad\quad = -2 \pm 2\sqrt{7}$$

but $0 \leqslant q_1 \leqslant 4$ $\qquad\qquad\qquad\qquad$ but $0 \leqslant q_3 \leqslant 4$

$$q_1 = -2 + 2\sqrt{3} = 1.464 \qquad\qquad q_3 = -2 + 2\sqrt{7} = 3.29$$

> The nth percentile, p_n, of a continuous random variable X having probability density function $f(x)$ is such that
>
> $$F(p_n) = P(X \leqslant p_n) = \int_{-\infty}^{p_n} f(x)\,dx = \frac{n}{100}$$

If $f(x)$ is defined on the range of values $a \leqslant x \leqslant b$ then

$$F(p_n) = P(X \leqslant p_n) = \frac{n}{100}$$

$$\Rightarrow \int_a^{p_n} f(x)\,dx = \frac{n}{100}$$

From this, you should be able to see that:

The lower quartile = 25th percentile.
The median = 50th percentile.
The upper quartile = 75th percentile.

Worked example 3.13

The continuous random variable X has the following cumulative distribution function:

$$F(x) = \begin{cases} 0 & x < 0 \\ \dfrac{1}{216}x^3 & 0 \leqslant x \leqslant 6 \\ 1 & x > 6 \end{cases}$$

Determine the values of the 10th and 90th percentiles of X.

Solution

10th percentile:

$$F(p_{10}) = \frac{1}{216}p_{10}{}^3 = \frac{10}{100} = 0.1$$

$$p_{10} = \sqrt[3]{21.6} = 2.78$$

90th percentile:

$$F(p_{90}) = \frac{1}{216}p_{90}{}^3 = \frac{90}{100} = 0.9$$

$$p_{90} = \sqrt[3]{194.4} = 5.79$$

EXERCISE 3C

1 The continuous random variable R has the cumulative distribution function

$$F(r) = \begin{cases} 0 & r < 2 \\ \frac{1}{20}(r^2 + 4r - 12) & 2 \leqslant r \leqslant 4 \\ 1 & r > 4 \end{cases}$$

(a) Sketch the graph of $F(r)$.

(b) Determine the value of:
   (i) the lower and upper quartiles of R,
   (ii) the median of R,
   (iii) the 90th percentile of R.

2 The continuous random variable Y has cumulative distribution function:

$$F(y) = \begin{cases} 0 & y < 5 \\ \frac{1}{10}(y - 5) & 5 \leqslant y \leqslant 15 \\ 1 & y > 15 \end{cases}$$

Determine the values of the median and the 80th percentile of Y.

3 The continuous random variable T has the following probability density function:

$$f(t) = \begin{cases} \frac{1}{3}(t - 1) & 1 \leqslant t < 2 \\ \frac{1}{15}(7 - t) & 2 \leqslant t < 7 \\ 0 & \text{otherwise.} \end{cases}$$

(a) Sketch the graph of f.

(b) Find the cumulative distribution function $F(t)$.

(c) Determine the values of the lower and upper quartiles, the median and the 10th percentile of T.

4 The continuous random variable X has the following probability density function:

$$f(x) = \begin{cases} \frac{1}{63}x^2 & 3 \leqslant x \leqslant 6 \\ 0 & \text{otherwise.} \end{cases}$$

(a) Sketch the graph of f.

(b) Find the cumulative distribution function, $F(x)$, and sketch its graph.

(c) Determine the values of:
  (i) the median, m,
  (ii) the 30th percentile, p_{30},
  (iii) the upper quartile, q_3.
(d) Calculate the value of $P(X > 5)$.

5 The continuous random variable X has the following probability density function:

$$f(x) = \begin{cases} \dfrac{1}{8} & -2 \leq x < 0 \\ \dfrac{1}{24}(2x + 3) & 0 \leq x \leq 3 \\ 0 & \text{otherwise.} \end{cases}$$

(a) Find the cumulative distribution function, $F(x)$.
(b) Determine the value of:
  (i) the median,
  (ii) the lower and upper quartiles,
  (iii) the 70th percentile.

3.5 The expectation of a continuous random variable

The expected value of a continuous random variable X, having a probability density function $f(x)$ is usually denoted by $E(X)$, where

$$E(X) = \int_{\text{all } x} x\, f(x)\, dx$$

Compared with the discrete case where $E(X) = \sum_{\text{all } x} x\, P(X = x)$.

Worked example 3.14

A random variable X, has the probability density function

$$f(x) = \begin{cases} 0.02x & 0 < x < 10 \\ 0 & \text{otherwise.} \end{cases}$$

Find $E(X)$.

Solution

$$E(X) = \int_0^{10} x\, f(x)\, dx$$
$$= \int_0^{10} 0.02x^2\, dx$$
$$= \left[\frac{0.02x^3}{3} \right]_0^{10}$$
$$= 6.67$$

$E(X)$ is the mean score we would expect to obtain if samples were repeatedly taken from this distribution.

3.6 The mean of a continuous random variable

The mean of a continuous random variable X is defined to be $E(X)$.

The mean of X is $E(X) = \int_{\text{all } x} x\, f(x)\, dx$.

We usually write the mean of X as $\mu = E(X)$.

Worked example 3.15

The continuous random variable X has a probability density function defined by

$$f(x) = \begin{cases} \dfrac{3}{4}(1 - x^2) & -1 \leqslant x \leqslant 1 \\ 0 & \text{otherwise.} \end{cases}$$

(a) Sketch the graph of f.

(b) Calculate the value of the mean of X.

Solution

(a)

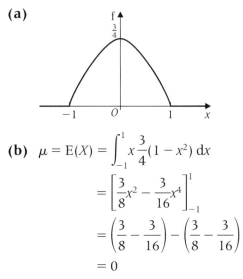

(b) $\mu = E(X) = \displaystyle\int_{-1}^{1} x\frac{3}{4}(1 - x^2)\, dx$

$= \left[\dfrac{3}{8}x^2 - \dfrac{3}{16}x^4 \right]_{-1}^{1}$

$= \left(\dfrac{3}{8} - \dfrac{3}{16} \right) - \left(\dfrac{3}{8} - \dfrac{3}{16} \right)$

$= 0$

This result could have been found by considering the symmetry of the graph of the p.d.f. f(x).

Worked example 3.16

The continuous random variable T has probability density function defined by

$$f(t) = \begin{cases} \dfrac{1}{12} & -8 \leqslant t < 0 \\ \dfrac{1}{48}(3t^2 + 4) & 0 \leqslant t < 2 \\ 0 & \text{otherwise.} \end{cases}$$

Calculate the value of the mean of T.

Solution

$$\mu = E(T) = \int_{\text{all } t} t\, f(t)\, dt$$

$$= \int_{-8}^{0} \frac{t}{12}\, dt + \int_{0}^{2} \frac{1}{48} t(3t^2 + 4)\, dt$$

$$= \left[\frac{t^2}{24} \right]_{-8}^{0} + \left[\frac{t^4}{64} + \frac{t^2}{24} \right]_{0}^{2}$$

$$= \left(0 - \frac{64}{24} \right) + \left(\frac{16}{64} + \frac{4}{24} - 0 \right)$$

$$= -2.25$$

EXERCISE 3D

1 The random variable Y has probability density function

$$f(y) = \begin{cases} \dfrac{1}{7} & -2 < y < 1 \\[2mm] \dfrac{2}{7} & 1 < y \leqslant 3 \\[2mm] 0 & \text{otherwise.} \end{cases}$$

   (a) Sketch the graph of f.

   (b) Find $E(Y)$.

2 The continuous random variable R has the following probability density function:

$$f(r) = \begin{cases} 0.25r & 1 \leqslant r \leqslant 3 \\ 0 & \text{otherwise.} \end{cases}$$

   (a) Sketch the graph of $f(r)$.

   (b) Find $E(R)$.

3 The continuous random variable X has the following probability density function:

$$f(x) = \begin{cases} kx(6 - x) & 2 \leqslant x \leqslant 6 \\ 0 & \text{otherwise.} \end{cases}$$

   (a) Show that $k = \dfrac{3}{80}$.

   (b) Sketch the graph of f.

   (c) Find $E(X)$.

4 A continuous random variable X has the probability density function:

$$f(x) = \begin{cases} \dfrac{3}{26}(x - 1)^2 & 2 \leqslant x \leqslant 4 \\[2mm] 0 & \text{otherwise.} \end{cases}$$

   (a) Sketch the graph of f.

   (b) Calculate $E(X)$.

3.7 The expectation of g(X)

The expectation of g(X), denoted by E[g(X)], where g(X) is any function of a continuous random variable X, having probability density function f(x), is defined to be:

$$E[g(X)] = \int_{\text{all } x} g(x)\, f(x)\, dx$$

Worked example 3.17

A continuous random variable X has the following probability density function:

$$f(x) = \begin{cases} \dfrac{1}{32}x & 0 \leqslant x \leqslant 8 \\ 0 & \text{otherwise.} \end{cases}$$

(a) Find:

    (i) E(X),

    (ii) E(3X).

(b) Hence show that E(3X) = 3 E(X).

Solution

(a) (i) $E(X) = \displaystyle\int_0^8 \frac{1}{32}x^2\, dx$

$$= \left[\frac{x^3}{96}\right]_0^8$$

$$= 5\frac{1}{3}$$

> Using $E(X) = \displaystyle\int_{\text{all } x} x\, f(x)\, dx.$

    (ii) $E(3X) = \displaystyle\int_0^8 \frac{3}{32}x^2\, dx$

$$= \left[\frac{x^3}{32}\right]_0^8$$

$$= 16$$

> Using $E[g(x)] = \displaystyle\int_{\text{all } x} g(x)\, f(x)\, dx$, where g(x) = 3x.

(b) $3E(X) = 3 \times 5\frac{1}{3} = 16 = E(3X)$

> In general E(aX) = aE(X), where a is a constant.

Worked example 3.18

The continuous random variable X has the following probability density function:

$$f(x) = \begin{cases} \dfrac{x}{6} & 0 \leqslant x < 2 \\ \dfrac{1}{3} & 2 \leqslant x < 4 \\ 0 & \text{otherwise.} \end{cases}$$

(a) Find:
   (i) $E(X)$,
   (ii) $E(12X + 5)$.

(b) Hence show that $E(12X + 5) = 12E(X) + 5$.

Solution

(a) (i) $E(X) = \int_0^2 \dfrac{x^2}{6}\,dx + \int_2^4 \dfrac{x}{3}\,dx$

$\qquad\qquad = \left[\dfrac{x^3}{18}\right]_0^2 + \left[\dfrac{x^2}{6}\right]_2^4$

$\qquad\qquad = \dfrac{4}{9} + \left(\dfrac{8}{3} - \dfrac{2}{3}\right)$

$\qquad\qquad = 2\tfrac{4}{9}$

   (ii) $E(12X + 5) = \int_0^2 (12x + 5)\dfrac{x}{6}\,dx + \int_2^4 \dfrac{1}{3}(12x + 5)\,dx$

$\qquad\qquad\qquad = \left[\dfrac{2x^3}{3} + \dfrac{5x^2}{12}\right]_0^2 + \left[2x^2 + \dfrac{5x}{3}\right]_2^4$

$\qquad\qquad\qquad = 7 + 27\tfrac{1}{3}$

$\qquad\qquad\qquad = 34\tfrac{1}{3}$

(b) $12E(X) + 5 = 12 \times 2\tfrac{4}{9} + 5 = 34\tfrac{1}{3} = E(12X + 5)$

> In general $E(aX + b) = aE(X) + b$, where a and b are constants.

Worked example 3.19

The continuous random variable X, has a probability density function given by

$$f(x) = \begin{cases} \dfrac{1}{64}x^3 & 0 \leqslant x \leqslant 4 \\ 0 & \text{otherwise.} \end{cases}$$

Find:

(a) (i) $E(X^2)$,
   (ii) $E\left(\dfrac{1}{X}\right)$.

(b) Hence write down the values of $E(6X^2)$ and $E\left(\dfrac{12}{X}\right)$.

Solution

(a) (i) $E(X^2) = \displaystyle\int_{\text{all }x} x^2\, f(x)\, dx$

$\qquad\qquad = \int_0^4 x^2\left(\dfrac{1}{64}x^3\right) dx = \int_0^4 \dfrac{1}{64}x^5\, dx$

$\qquad\qquad = \left[\dfrac{x^6}{384}\right]_0^4$

$\qquad\qquad = 10\tfrac{2}{3}$

(ii) $E\left(\dfrac{1}{X}\right) = \displaystyle\int_{\text{all } x} \dfrac{1}{X} f(x) \, dx$

$$= \int_0^4 \dfrac{1}{x}\left(\dfrac{x^3}{64}\right) dx = \int_0^4 \dfrac{x^2}{64} \, dx$$

$$= \left[\dfrac{x^3}{192}\right]_0^4$$

$$= \dfrac{1}{3}$$

(b) $E(6X^2) = 6E(X^2) = 6 \times 10\tfrac{2}{3} = 64$

$$E\left(\dfrac{12}{X}\right) = 12\, E\left(\dfrac{1}{X}\right) = 12 \times \dfrac{1}{3} = 4$$

EXERCISE 3E

1 The continuous random variable R has the following probability density function:

$$f(r) = \begin{cases} \dfrac{3}{4}r\,(2-r) & 0 \leqslant r \leqslant 2 \\ 0 & \text{otherwise.} \end{cases}$$

(a) Find:

(i) $E(R^2)$,

(ii) $E(R^{-1})$.

(b) Hence find $E(10R^2)$ and $E(100R^{-1})$.

2 The continuous random variable X has the probability density function defined by

$$f(x) = \begin{cases} \dfrac{1}{18} & -6 < x \leqslant 0 \\ \dfrac{1}{18}(x^2+1) & 0 < x \leqslant 3 \\ 0 & \text{otherwise.} \end{cases}$$

(a) Determine the value of $E(X^2)$.

(b) Hence find $E(10X^2 + 1)$.

3 The continuous random variable Y has the following probability density function:

$$f(y) = \begin{cases} \dfrac{1}{6}y^2(3-y) & 1 \leqslant y \leqslant 3 \\ 0 & \text{otherwise.} \end{cases}$$

(a) Determine the values of $E(Y^2)$ and $E(Y^{-2})$.

(b) Hence find $E(90Y^2)$ and $E\left(\dfrac{90}{Y^2}\right)$.

3.8 The variance and standard deviation of a continuous random variable

The variance of a continuous random variable X is defined by:

$$\begin{aligned}\mathrm{Var}(X) &= \mathrm{E}(X - \mathrm{E}(X))^2 \\ &= \mathrm{E}(X^2) - [\mathrm{E}(X)]^2 \\ &= \mathrm{E}(X^2) - \mu^2\end{aligned}$$

The variance of X is usually denoted by $\mathrm{Var}(X) = \sigma^2$.

3

The standard deviation of a continuous random variable X is defined by:

$$\text{Standard deviation of } X = \sqrt{\mathrm{Var}(X)} = \sqrt{\mathrm{E}(X^2) - \mu^2}.$$

The standard deviation is usually denoted by σ.

Worked example 3.20

The random variable X has the following probability density function:

$$\mathrm{f}(x) = \begin{cases} 0.02x & 0 < x < 10 \\ 0 & \text{otherwise.} \end{cases}$$

Find the variance and standard deviation of X, given that $\mathrm{E}(X) = 6\frac{2}{3}$.

Solution

$$\begin{aligned}\mathrm{E}(X^2) &= \int_0^{10} 0.02x^3 \, \mathrm{d}x \\ &= \left[\frac{0.02x^4}{4}\right]_0^{10} = 50\end{aligned}$$

Using $\mathrm{E}(\mathrm{g}(X)) = \int_{\text{all } x} \mathrm{g}(x)\,\mathrm{f}(x)\,\mathrm{d}x$, where $\mathrm{g}(x) = x^2$.

$$\begin{aligned}\sigma^2 = \mathrm{Var}(X) &= \mathrm{E}(X^2) - [\mathrm{E}(X)]^2 \\ &= 50 - (6\tfrac{2}{3})^2 \\ &= 5\tfrac{5}{9} = 5.556\end{aligned}$$

$$\begin{aligned}\sigma &= \text{standard deviation of } X \\ &= \sqrt{5\tfrac{5}{9}} \\ &= 2.357\end{aligned}$$

Worked example 3.21

A continuous random variable Y has probability density function

$$\mathrm{f}(y) = \begin{cases} ky & 1 < y < 3 \\ 0 & \text{otherwise,} \end{cases}$$

where k is a constant.

(a) Find the value of k.

(b) Find the expectation and standard deviation of Y.

Solution

(a)

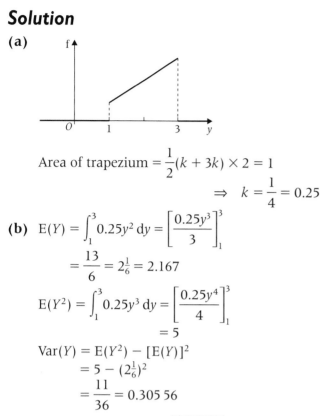

$$\text{Area of trapezium} = \frac{1}{2}(k + 3k) \times 2 = 1$$
$$\Rightarrow \quad k = \frac{1}{4} = 0.25$$

You could use integration to find the value of k.

$$\int_1^3 ky\,dy = \left[\frac{ky^2}{2}\right]_1^3 = 1$$
$$4k = 1$$
$$k = \frac{1}{4} = 0.25$$

(b) $E(Y) = \int_1^3 0.25y^2\,dy = \left[\frac{0.25y^3}{3}\right]_1^3$

$$= \frac{13}{6} = 2\frac{1}{6} = 2.167$$

$$E(Y^2) = \int_1^3 0.25y^3\,dy = \left[\frac{0.25y^4}{4}\right]_1^3$$
$$= 5$$

$$Var(Y) = E(Y^2) - [E(Y)]^2$$
$$= 5 - (2\frac{1}{6})^2$$
$$= \frac{11}{36} = 0.305\,56$$

$\therefore$ Standard deviation $= \sqrt{0.305\,56} = 0.5528.$

Use $E(Y) = \int_1^3 y\,f(y)\,dy.$

Use $E(Y^2) = \int_1^3 y^2\,f(y)\,dy.$

Standard deviation $= \sqrt{\text{variance}}.$

Worked example 3.22

A charity group raises funds by collecting waste paper. A full skip will contain an amount, X, of other materials such as plastic bags and rubber bands. X may be regarded as a random variable with probability density function

$$f(x) = \begin{cases} \dfrac{2}{9}(x-1)(4-x) & 1 < x < 4 \\ 0 & \text{otherwise.} \end{cases}$$

(All numerical values are in units of 100 kg.)

(a) Find the mean and standard deviation of X.

(b) Find the probability that X exceeds 3.5.

A full skip may normally be sold for £250 but if X exceeds 3.5 only £125 will be paid.

(c) Find, to the nearest £, the expected value of a full skip.

Alternatively the paper may be sorted before being placed in the skip. This will ensure a very low value of X and a full skip may then be sold for £310. However, the effort put into sorting, means that 25 per cent fewer skips will be sold.

(d) Advise the charity on whether or not they should sort the paper.

Solution

(a) $\mu = \mathrm{E}(X) = \int_1^4 \frac{2}{9} x(x-1)(4-x)\,\mathrm{d}x$

$\qquad = \frac{2}{9} \int_1^4 (-x^3 + 5x^2 - 4x)\,\mathrm{d}x$

$\qquad = \frac{2}{9} \left[-\frac{x^4}{4} + \frac{5x^3}{3} - 2x^2 \right]_1^4$

$\qquad = 2.5$

$\mathrm{E}(X^2) = \frac{2}{9} \int_1^4 (-x^4 + 5x^3 - 4x^2)\,\mathrm{d}x$

$\qquad = \frac{2}{9} \left[-\frac{x^5}{5} + \frac{5x^4}{4} - \frac{4x^3}{3} \right]_1^4$

$\qquad = 6.7$

$\therefore \quad \mathrm{Var}(X) = 6.7 - (2.5)^2 = 0.45$

$\Rightarrow \quad$ standard deviation of $X = \sqrt{0.45} = 0.671$.

(b) The probability that X exceeds 3.5 is given by

$$P(X > 3.5) = \frac{2}{9} \int_{3.5}^4 (-x^2 + 5x - 4)\,\mathrm{d}x$$

$$= \frac{2}{9} \left[-\frac{x^3}{3} + \frac{5x^2}{2} - 4x \right]_{3.5}^4$$

$$= 0.0741$$

(c) Expected value of a full skip is

$$£125 \times 0.0741 + £250 \times (1 - 0.0741) = £241.$$

(d) If skips are sorted, only 75% of the skips would be sold at £310, i.e. $0.75 \times £310 = £232.50$.

This is less than £241 and implies that the charity will make more money if the paper is **not** sorted.

EXERCISE 3F

1 A continuous random variable X, has a probability density function given by

$$f(x) = \begin{cases} kx & 1 < x < 2 \\ 0 & \text{otherwise,} \end{cases}$$

where k is a constant.

(a) Find the value of k.

(b) Find: **(i)** $\mathrm{E}(X)$, **(ii)** $\mathrm{E}(X^2)$.

(c) Hence find the variance and standard deviation of X.

2 A continuous random variable Y has the following probability density function:

$$f(y) = \begin{cases} 0.375y^2 & 0 < y < 2 \\ 0 & \text{otherwise.} \end{cases}$$

   (a) Find the mean, μ, and standard deviation, σ, of Y.

   (b) Determine $P(Y < \mu)$.

3 A temporary roundabout is installed at a crossroads. The time, T minutes, which vehicles have to wait before entering the crossroads has a probability density function

$$f(t) = \begin{cases} 0.8 - 0.32t & 0 < t < 2.5 \\ 0 & \text{otherwise.} \end{cases}$$

   Find the mean and standard deviation of T.

4 A continuous random variable X, has the probability density function

$$f(x) = \begin{cases} 0.15x(x + 2) & 0 < x < 2 \\ 0 & \text{otherwise.} \end{cases}$$

   (a) Calculate the mean and variance of X.

   (b) Determine $P(X \leqslant 1)$.

3.9 The variance of a simple function of a continuous random variable

> $\text{Var}(a) = 0$
> $\text{Var}(aX) = a^2 \, \text{Var}(X)$
> $\text{Var}(aX + b) = a^2 \, \text{Var}(X),$
>
> where a and b are constants.

The standard deviation of X is found by evaluating $\sqrt{\text{Var}(X)}$.

Worked example 3.23

The continuous random variable X has a probability density function defined by

$$f(x) = \begin{cases} \dfrac{1}{16}(x + 1) & 1 \leqslant x \leqslant 5 \\ 0 & \text{otherwise.} \end{cases}$$

(a) Sketch the graph of f.

(b) Find: **(i)** $E(X)$ and **(ii)** $E(X^2)$.

(c) Hence show that $\text{Var}(X) = 1\frac{2}{9}$.

(d) Find $\text{Var}(3X + 4)$.

Solution

(a)

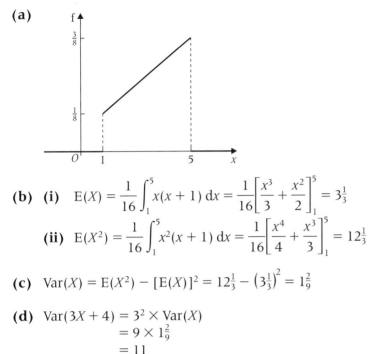

(b) (i) $E(X) = \dfrac{1}{16}\displaystyle\int_1^5 x(x+1)\,dx = \dfrac{1}{16}\left[\dfrac{x^3}{3} + \dfrac{x^2}{2}\right]_1^5 = 3\tfrac{1}{3}$

Using $E(X) = \displaystyle\int_{\text{all }x} x\,f(x)\,dx$.

(ii) $E(X^2) = \dfrac{1}{16}\displaystyle\int_1^5 x^2(x+1)\,dx = \dfrac{1}{16}\left[\dfrac{x^4}{4} + \dfrac{x^3}{3}\right]_1^5 = 12\tfrac{1}{3}$

Using $E(X^2) = \displaystyle\int_{\text{all }x} x^2\,f(x)\,dx$.

(c) $Var(X) = E(X^2) - [E(X)]^2 = 12\tfrac{1}{3} - \left(3\tfrac{1}{3}\right)^2 = 1\tfrac{2}{9}$

(d) $Var(3X+4) = 3^2 \times Var(X)$
$= 9 \times 1\tfrac{2}{9}$
$= 11$

Using $Var(aX + b) = a^2 \times Var(X)$.

Worked example 3.24

A continuous random variable X has the following probability density function

$$f(x) = \begin{cases} \dfrac{1}{9}x^2 & 0 \leqslant x \leqslant 3 \\ 0 & \text{otherwise.} \end{cases}$$

(a) Calculate $E(X)$ and $E(X^2)$.

(b) Hence calculate $Var(X)$.

(c) Show that $Var(4X - 1) = 5.4$.

Solution

(a) $E(X) = \displaystyle\int_0^3 x\,\dfrac{1}{9}x^2\,dx = \left[\dfrac{1}{36}x^4\right]_0^3 = 2.25$

$E(X^2) = \displaystyle\int_0^3 x^2\,\dfrac{1}{9}x^2\,dx = \left[\dfrac{1}{45}x^5\right]_0^3 = 5.4$

(b) $Var(X) = 5.4 - (2.25)^2 = \dfrac{27}{80} = 0.3375$

(c) $Var(4X - 1) = 4^2 \times Var(X) = 16 \times 0.3375 = 5.4$

Worked example 3.25

A continuous random variable X has probability density function

$$f(x) = \begin{cases} \dfrac{1}{2}x & 0 \leqslant x < 1 \\ \dfrac{1}{2} & 1 \leqslant x \leqslant 2.5 \\ 0 & \text{otherwise.} \end{cases}$$

(a) Calculate the mean and variance of X.

(b) Hence find $\text{Var}(4X + 3)$.

Solution

(a) $E(X) = \displaystyle\int_0^1 \dfrac{1}{2}x^2 \, dx + \int_1^{2.5} \dfrac{1}{2}x \, dx$

$$= \left[\dfrac{x^3}{3}\right]_0^1 + \left[\dfrac{x^2}{4}\right]_1^{2.5}$$

$$= 1.479$$

$E(X^2) = \displaystyle\int_0^1 \dfrac{1}{2}x^3 \, dx + \int_1^{2.5} \dfrac{1}{2}x^2 \, dx$

$$= \left[\dfrac{x^4}{8}\right]_0^1 + \left[\dfrac{x^3}{6}\right]_1^{2.5}$$

$$= 2.5625$$

$\text{Var}(X) = E(X^2) - [E(X)]^2$
$\quad\quad\quad = 2.5625 - (1.479)^2$
$\quad\quad\quad = 0.375$

(b) $\text{Var}(4X + 3) = 4^2 \times \text{Var}(X)$
$\quad\quad\quad\quad\quad\quad = 16 \times 0.375$
$\quad\quad\quad\quad\quad\quad = 6$

Worked example 3.26

The continuous random variable X has the probability density function

$$f(x) = \begin{cases} \dfrac{4}{81}x^2(4 - x) & 1 \leqslant x \leqslant 4 \\ 0 & \text{otherwise.} \end{cases}$$

(a) Calculate: **(i)** $E(X^{-1})$, **(ii)** $E(X^{-2})$.

(b) Hence show that $\text{Var}(X^{-1}) = \dfrac{14}{81}$.

(c) Deduce the value of $\text{Var}\left(\dfrac{9}{X}\right)$.

Solution

(a) (i)
$$E(X^{-1}) = \int_1^4 x^{-1} \frac{4}{81} x^2 (4-x) \, dx$$
$$= \frac{4}{81} \int_1^4 x(4-x) \, dx$$
$$= \frac{4}{81} \left[2x^2 - \frac{x^3}{3} \right]_1^4$$
$$= \frac{4}{9}$$

(ii)
$$E(X^{-2}) = \int_1^4 x^{-2} \frac{4}{81} x^2 (4-x) \, dx$$
$$= \frac{4}{81} \int_1^4 (4-x) \, dx$$
$$= \frac{4}{81} \left[4x - \frac{x^2}{2} \right]_1^4$$
$$= \frac{2}{9}$$

(b) $\text{Var}(X^{-1}) = \dfrac{2}{9} - \left(\dfrac{4}{9}\right)^2 = \dfrac{2}{81}$

(c) $\text{Var}\left(\dfrac{9}{X}\right) = 9^2 \times \text{Var}(X) = 81 \times \dfrac{2}{81} = 2$

EXERCISE 3G

1 The continuous random variable X has the probability density function

$$f(x) = \begin{cases} \dfrac{1}{24}(x+3) & 1 \leqslant x \leqslant 5 \\ 0 & \text{otherwise.} \end{cases}$$

(a) Calculate $E(X)$ and $E(X^2)$.

(b) Hence find the value of the variance of X.

(c) Deduce the values of: **(i)** $\text{Var}(10X)$, **(ii)** $\text{Var}(10X - 7)$.

2 The continuous random variable X has the following probability density function:

$$f(x) = \begin{cases} \dfrac{1}{960} x^3 & 4 \leqslant x \leqslant 8 \\ 0 & \text{otherwise.} \end{cases}$$

(a) Calculate the value of $E\left(\dfrac{1}{X^2}\right)$.

(b) Given that $E\left(\dfrac{1}{X}\right) = \dfrac{7}{45}$, find the value of $\text{Var}\left(\dfrac{1}{X}\right)$.

(c) Hence show that $\text{Var}(100X^{-1}) \simeq 8$.

3 The continuous random variable R has the probability density function

$$f(r) = \begin{cases} kr^4 & 1 \leqslant r \leqslant 2 \\ 0 & \text{otherwise,} \end{cases}$$

where k is a constant.

(a) Show that $k = \dfrac{5}{31}$.

(b) Calculate $E(R^{-1})$ and $E(R^{-2})$.

(c) Hence find: **(i)** $\text{Var}(R^{-1})$ and **(ii)** $\text{Var}(10R^{-1})$.

4 The continuous random variable X has the following probability density function

$$f(x) = \begin{cases} \dfrac{3}{23}x^2 & 1 \leqslant x < 2 \\ \dfrac{3}{23}x(4-x) & 2 \leqslant x \leqslant 4 \\ 0 & \text{otherwise.} \end{cases}$$

(a) Sketch the graph of f.

(b) Calculate $E(X)$ and $E(X^2)$.

(c) Find $\text{Var}(X)$ and hence write down the value of $\text{Var}(10X + 3)$.

5 The continuous random variable Y has the probability density function

$$f(y) = \begin{cases} \dfrac{3}{79}y^2 & 1 \leqslant y < 2 \\ \dfrac{3}{316}y^2(6-y) & 2 \leqslant y \leqslant 6 \\ 0 & \text{otherwise.} \end{cases}$$

(a) Sketch the graph of f.

(b) Find $E\left(\dfrac{1}{Y}\right)$ and $E\left(\dfrac{1}{Y^2}\right)$.

(c) Hence calculate the value of $\text{Var}\left(\dfrac{1}{Y}\right)$, giving your answer to two significant figures.

(d) Write down the value of:

(i) $\text{Var}\left(\dfrac{10}{Y}\right)$,

(ii) $\text{Var}\left(\dfrac{10}{Y} + 2\right)$.

3.10 The rectangular distribution

The random variable X having probability density function

$$f(x) = \begin{cases} \dfrac{1}{b - a} & a < x < b \\ 0 & \text{otherwise,} \end{cases}$$

where a and b are constants, is said to follow a rectangular distribution.

This distribution is sometimes called the continuous uniform distribution.

The **mean** and **variance** of a continuous random variable X, which follows a **rectangular** distribution, are

$$E(X) = \frac{1}{2}(a + b)$$

$$Var(X) = \frac{1}{12}(b - a)^2$$

Worked example 3.27

Prove that $E(X) = \frac{1}{2}(a + b)$, where the random variable X follows a rectangular distribution.

Solution

$$E(X) = \int_a^b x \frac{1}{b - a} \, dx$$

$$= \frac{1}{b - a} \int_a^b x \, dx$$

$$= \frac{1}{b - a} \left[\frac{x^2}{2} \right]_a^b$$

$$= \frac{1}{b - a} \left[\frac{b^2 - a^2}{2} \right]$$

$$= \frac{1}{b - a} \left[\frac{(b - a)(b + a)}{2} \right]$$

$$= \frac{1}{2}(a + b)$$

Using $E(X) = \int_{\text{all } x} x\, f(x)\, dx.$

Worked example 3.28

Prove that, for a random variable X which follows a rectangular distribution, $Var(X) = \frac{1}{12}(b - a)^2$.

Solution

$$E(X^2) = \int_a^b x^2 \frac{1}{b - a} \, dx$$

$$= \frac{1}{b - a} \int_a^b x^2 \, dx$$

$$= \frac{1}{b - a} \left[\frac{x^3}{3} \right]_a^b$$

$$= \frac{1}{b - a} \left[\frac{b^3 - a^3}{3} \right]$$

$$= \frac{1}{b - a} \left[\frac{(b - a)(b^2 + ab + a^2)}{3} \right]$$

$$= \frac{1}{3}(b^2 + ab + a^2)$$

Using $E[g(X)] = \int_{\text{all } x} g(x)\, f(x)\, dx,$
where $g(x) = x^2$

$$\text{Var}(X) = \text{E}(X^2) - [\text{E}(X)]^2$$

$$= \frac{1}{3}(b^2 + ab + a^2) - \left[\frac{1}{2}(a + b)\right]^2$$

$$= \frac{1}{3}(b^2 + ab + a^2) - \frac{1}{4}(a^2 + 2ab + b^2)$$

$$= \frac{1}{12}[4(b^2 + ab + a^2) - 3(a^2 + 2ab + b^2)]$$

$$= \frac{1}{12}[b^2 - 2ab + a^2]$$

$$= \frac{1}{12}(b - a)^2$$

> You may be asked to prove $\text{E}(X) = \frac{1}{2}(a + b)$ and $\text{Var}(X) = \frac{1}{12}(b - a)^2$ as part of the examination.

Worked example 3.29

The continuous random variable X has the probability density function

$$f(x) = \begin{cases} \dfrac{1}{6} & -2 \leqslant x \leqslant 4 \\ 0 & \text{otherwise.} \end{cases}$$

(a) Sketch the graph of f.

(b) Find the mean and variance of X.

(c) Calculate $\text{P}(|X| < 1)$.

Solution

(a)

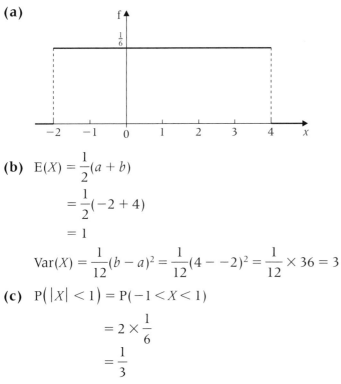

(b) $\text{E}(X) = \dfrac{1}{2}(a + b)$

$$= \frac{1}{2}(-2 + 4)$$

$$= 1$$

$$\text{Var}(X) = \frac{1}{12}(b - a)^2 = \frac{1}{12}(4 - -2)^2 = \frac{1}{12} \times 36 = 3$$

(c) $\text{P}(|X| < 1) = \text{P}(-1 < X < 1)$

$$= 2 \times \frac{1}{6}$$

$$= \frac{1}{3}$$

Worked example 3.30

The continuous random variable R, has the following probability density function

$$f(r) = \begin{cases} k & 1 < r < 6 \\ 0 & \text{otherwise,} \end{cases}$$

where k is a constant.

(a) Show that $k = \dfrac{1}{5}$.

(b) Find $E(R)$.

(c) Show that $\sigma = \dfrac{5\sqrt{3}}{6}$.

(d) Find $P(R < 3)$.

Solution

(a) $k = \dfrac{1}{b-a} = \dfrac{1}{6-1} = \dfrac{1}{5}$

(b) $E(R) = \dfrac{1}{2}(a+b) = \dfrac{1}{2}(6+1) = 3.5$

(c) $\sigma^2 = \text{Var}(R) = \dfrac{1}{12}(b-a)^2$

$$= \dfrac{1}{12} \times 5^2$$

$$= \dfrac{25}{12}$$

∴ The standard deviation of R is given by:

$$\sigma = \sqrt{\dfrac{25}{12}} = \dfrac{5}{\sqrt{12}} = \dfrac{5}{2\sqrt{3}} \quad \Rightarrow \quad \sigma = \dfrac{5\sqrt{3}}{6}$$

(d) $P(R < 3) = 2 \times \dfrac{1}{5} = 0.4$

EXERCISE 3H

1 Heights of children are measured to the nearest centimetre. The rounding error may be regarded as a random variable X, with probability density function

$$f(x) = \begin{cases} 1 & -0.5 < x < 0.5 \\ 0 & \text{otherwise.} \end{cases}$$

(a) Find the mean and standard deviation of X.

(b) Show that, for a particular measurement, the probability that the **magnitude** of the rounding error is less than 0.05 cm is 0.1.

2 The random variable X, has a probability density function defined by

$$f(x) = \begin{cases} k & 4 < x < 9 \\ 0 & \text{otherwise,} \end{cases}$$

where k is a constant.

Find:

(a) k,

(b) the mean and variance of X,

(c) the standard deviation of X.

3 The random variable Y has the probability density function

$$f(y) = \begin{cases} k & 0 < y < 10 \\ 0 & \text{otherwise.} \end{cases}$$

Find:

(a) k,

(b) $P(Y = 4)$,

(c) $P(4 < Y < 7)$,

(d) the mean and standard deviation of Y.

4 A continuous random variable X, has a probability density function

$$f(x) = \begin{cases} k & -1 < x < 4 \\ 0 & \text{otherwise.} \end{cases}$$

Find:

(a) k,

(b) $P(X = 1)$,

(c) $P(0 < X < 1)$,

(d) the mean, variance and standard deviation of X.

5 A technique for measuring the density of a silicon compound results in an error which may be modelled by the random variable X, with probability density function

$$f(x) = \begin{cases} k & -0.04 < x < 0.04 \\ 0 & \text{otherwise.} \end{cases}$$

Find:

(a) the value of k,

(b) the mean and standard deviation of X,

(c) the probability that the error is between -0.03 and 0.01,

(d) the probability that the magnitude of the error is greater than 0.035.

MIXED EXERCISE

1 The error, in grams, made by a greengrocer's scales may be modelled by the random variable X, with probability density function

$$f(x) = \begin{cases} 0.1 & -3 < x < 7 \\ 0 & \text{otherwise.} \end{cases}$$

Find the probability that:

(a) the error is positive,

(b) the magnitude of an error exceeds 2 grams, $\left[\text{i.e. } P(|X| > 2)\right]$,

(c) the magnitude of an error is less than 4 grams, $\left[\text{i.e. } P(|X| < 4)\right]$.

2 In an attempt to economise on her telephone bill, Debbie times her calls and ensures that they never last longer than four minutes. The length of her calls, T minutes, may be regarded as a random variable with probability density function

$$f(t) = \begin{cases} kt & 0 < t < 4 \\ 0 & \text{otherwise,} \end{cases}$$

where k is a constant.

(a) Show that $k = 0.125$.

(b) Find the mean and standard deviation of T.

(c) Find the probability that a call lasts for between three and four minutes.

3 The time, T hours, taken by any member of a group of friends to complete a run for charity can be modelled by the following probability density function:

$$f(t) = \begin{cases} \dfrac{1}{90}t^2 & 3 \leqslant t \leqslant 6 \\[2mm] 2 - \dfrac{4}{15}t & 6 \leqslant t \leqslant 7.5 \\[2mm] 0 & \text{otherwise.} \end{cases}$$

(a) **(i)** Show that the probability that a member of the group selected at random takes at least six hours to complete the run is 0.3.

   (ii) Evaluate the median time taken to complete the run.

(b) Calculate the mean time taken to complete the run.   [A]

4 The errors, in grams, made by a butcher's scales may be modelled by a random variable Y, with probability density function

$$f(y) = \begin{cases} k & -6 < y < 14 \\ 0 & \text{otherwise.} \end{cases}$$

(a) Show that $k = 0.05$.

(b) Write down the value of the mean of Y.

(c) **(i)** Use integration to find $E(Y^2)$.
   (ii) Hence, or otherwise, find the standard deviation of Y.

(d) **(i)** Find the probability that an error is negative.
   (ii) Show that the probability of an error having a magnitude less than 2 is 0.2.
   (iii) Find the probability that the magnitude of an error is greater than 4. [A]

5 The probability density function, $f(x)$, for the difference in millimetres from the nominal length of metal rods cut by a faulty machine, is shown by the following graph.

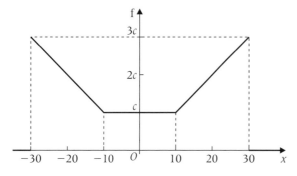

(a) Show that the value of the constant c, as used on the vertical axis, is 0.01.

(b) Determine:
   (i) $P(X > 20)$,
   (ii) $P(X < 10)$. [A]

6 A video recorder clock is corrected automatically by a signal every 24 hours. The deviation, T seconds, of the clock from the correct time, immediately prior to correction, may be modelled by the following probability density function:

$$f(t) = \begin{cases} \dfrac{1}{5b} & -b < t < 4b \\ 0 & \text{otherwise.} \end{cases}$$

(a) Given that $E(T) = 9$, show that $b = 6$.

(b) Hence calculate $P(T < 15)$. [A]

7 All consultations with an optician are by appointment. The time, T minutes, by which an appointment is delayed has the following probability density function:

$$f(t) = \begin{cases} \dfrac{1}{54}t^2 & 0 \leqslant t < 3 \\[2mm] \dfrac{1}{6} & 3 \leqslant t < 6 \\[2mm] \dfrac{1}{24}(10 - t) & 6 \leqslant t < 10 \\[2mm] 0 & \text{otherwise.} \end{cases}$$

(a) Sketch the graph of f.

(b) Use your graph to find:
  (i) $P(3 \leqslant T < 6)$,
  (ii) $P(T > 6)$,
  (iii) the lower quartile.

(c) What is the proportion of appointments delayed by less than 2 minutes? [A]

8 In an attempt to economise on her mobile phone costs, Tara times her calls to ensure that no call lasts longer than six minutes. The resultant lengths of calls, T minutes, may be regarded as a continuous random variable with the following probability density function.

$$f(t) = \begin{cases} k(2t + 3) & 1 \leqslant t \leqslant 6 \\ 0 & \text{otherwise.} \end{cases}$$

(a) Sketch the graph of f.

(b) By considering your sketch, or otherwise, show that the value of the constant k is $\dfrac{1}{50}$.

(c) Determine the probability that the length of a call is:
  (i) at most three minutes,
  (ii) between two and five minutes. [A]

9 The random variable X has the probability density function:

$$f(x) = \begin{cases} px + q & 0 < x < 1 \\ 0 & \text{otherwise,} \end{cases}$$

where p and q are constants.

(a) Show that $\dfrac{1}{2}p + q = 1$.

(b) Find the mean of X in terms of p and q.

(c) The mean of X is 0.6. Show that $p = 1.2$ and find the value of q.

(d) Find the standard deviation of X. [A]

10 The continuous random variable X has a rectangular
distribution on the interval (a, b), where $0 < a < b$.

  (a) Given that the mean, μ, is equal to 21 and that the
variance, σ^2, is equal to 27, prove that $a = 12$ and $b = 30$.

  (b) Hence determine:
    (i) $\mathrm{P}(5 < X < 20)$,
    (ii) $\mathrm{P}\left(X < \mu - \dfrac{\sigma\sqrt{3}}{2}\right)$.   [A]

11 Customers at an Internet café pay a flat fee which entitles
them to use a terminal for up to two hours. The actual
amount of time, in hours, that a customer uses a terminal,
may be modelled by the random variable T, with the
probability density function

$$\mathrm{f}(t) = \begin{cases} kt & 0 < t < 2 \\ 0 & \text{otherwise.} \end{cases}$$

  (a) Show that the value of the constant k is 0.5.

  (b) Find the mean and the standard deviation of T.

  (c) Find the probability that a customer spends less than
one hour using a terminal.   [A]

12 The random variable X, has the probability density function
defined by:

$$\mathrm{f}(x) = \begin{cases} k(9 - 2x) & 0 < x < 3 \\ 0 & \text{otherwise.} \end{cases}$$

  (a) Verify that $k = \dfrac{1}{18}$.

  (b) Find the mean and the standard deviation of X.   [A]

13 The time, H hours, to carry out a 24 000 miles service on a car
may be regarded as a continuous random variable with the
following probability density function:

$$\mathrm{f}(h) = \begin{cases} c(7 - 2h) & 1 \leqslant h \leqslant 3 \\ 0 & \text{otherwise.} \end{cases}$$

  (a) Sketch the graph of f.

  (b) By considering your sketch, or otherwise, show that the
value of the constant c is $\dfrac{1}{6}$.

  (c) Calculate the probability that a service takes more than
2 hours.   [A]

14 The total amount of added time, S minutes, played per match by a team in a local football league is distributed with the probability density function given by:

$$f(s) = \begin{cases} ks^2 & 1 \leqslant s \leqslant 5 \\ 0 & \text{otherwise.} \end{cases}$$

(a) Show that the exact value of the constant k is $\dfrac{3}{124}$.

(b) Determine $P(S < 3)$.

(c) Calculate the median amount of added time played per match, giving your answer to the nearest minute.   [A]

15 A darts player practises by throwing darts at the bull. The distance, X millimetres, by which he misses the bull may be modelled by the following probability density function:

$$f(x) = \begin{cases} 0.006x(10 - x) & 0 \leqslant x < 10 \\ 0 & \text{otherwise.} \end{cases}$$

(a) Calculate the probability that he misses the bull by more than 6 mm.

(b) Write down the value of $E(X)$.   [A]

Key point summary

1 $F(x) = P(x \leqslant x) = \displaystyle\int_{-\infty}^{x} f(x)\,dx$ *p51*

$$\frac{d}{dx}F(x) = f(x)$$

$F(x)$ is called the cumulative distribution function.

2 $F(q_1) = 0.25$ and $F(q_3) = 0.75$,   *p58*
where q_1 is the lower quartile
and q_3 is the upper quartile.

3 The nth percentile p_n is such that   *p59*

$$F(p_n) = \frac{n}{100}.$$

4 For a continuous random variable X,   *p61*

$$E(X) = \int_{\text{all } x} x\,f(x)\,dx.$$

5 If $g(x)$ is any function of a continuous random variable X,   *p64*

$$E[g(x)] = \int_{\text{all } x} g(x)\,f(x)\,dx.$$

6 Mean $= \mu = E(X)$.   *p62, 67*
Variance $= \sigma^2 = \text{Var}(X) = E(X^2) - [E(X)]^2$.
The standard deviation of $X = \sigma = \sqrt{\text{Var}(X)}$.

7 For a continuous random variable X, and   *p64, 65, 70*
constants a and b,

$\mathrm{E}(a) = a$ $\mathrm{Var}(a) = 0$

$\mathrm{E}(aX) = a\mathrm{E}(X)$ $\mathrm{Var}(aX) = a^2\,\mathrm{Var}(X)$

$\mathrm{E}(aX + b) = a\mathrm{E}(X) + b$ $\mathrm{Var}(aX + b) = a^2\,\mathrm{Var}(X)$

8 The probability density function of a continuous   *p74, 75*
random variable X, that follows a rectangular
distribution is of the form

$$f(x) = \begin{cases} \dfrac{1}{b-a} & a < x < b \\ 0 & \text{otherwise,} \end{cases}$$

where a and b are constants.

$$\mu = \mathrm{E}(X) = \frac{1}{2}(a + b)$$

$$\sigma^2 = \mathrm{Var}(X) = \frac{1}{12}(b - a)^2$$

Test yourself | What to review

1 A continuous random variable X has the probability density   *Section 3.2*
function defined by

$$f(x) = \begin{cases} kx^2 & 0 < x < 2 \\ 0 & \text{otherwise.} \end{cases}$$

(a) Show that $k = \dfrac{3}{8}$, and sketch the graph of f.

(b) Find: (i) $\mathrm{P}(X > 0.5)$ and (ii) $\mathrm{P}(X = 1)$.

---

2 The continuous random variable R has a probability density   *Section 3.3*
function defined by

$$f(r) = \begin{cases} \dfrac{1}{2}r & 0 < r < 2 \\ 0 & \text{otherwise.} \end{cases}$$

(a) Find the cumulative distribution function, $\mathrm{F}(r)$.

(b) Hence find $\mathrm{P}(R \leqslant 1.5)$.

---

3 The continuous random variable S has the following
cumulative distribution function, $\mathrm{F}(s)$:

$$\mathrm{F}(s) = \begin{cases} 0 & s < 0 \\ \dfrac{8}{27}s^3 & 0 \leqslant s \leqslant 1.5 \\ 1 & s > 1.5. \end{cases}$$

(a) Calculate the value of the median of S.   *Section 3.4*

(b) Find the value of the 90th percentile.

(c) Find $\mathrm{P}(S < 1.2)$ *Section 3.3*

4 The continuous random variable X has the probability density function

$$f(x) = \begin{cases} \dfrac{8}{9}x^2 & 0 \leqslant x \leqslant 1.5 \\ 0 & \text{otherwise.} \end{cases}$$

(a) Calculate $E(X)$. *Sections 3.5 and 3.6*

(b) Hence find: **(i)** $E(4X)$ and **(ii)** $E(4X + 5)$. *Section 3.7*

---

5 The continuous random variable Y has the following probability density function:

$$f(y) = \begin{cases} \dfrac{4}{625}y^3 & 0 \leqslant y \leqslant 5 \\ 0 & \text{otherwise.} \end{cases}$$

(a) Calculate: **(i)** $E(Y)$ and **(ii)** $E(Y^2)$. *Section 3.7*

(b) Hence find the variance and standard deviation of Y. *Sections 3.6 and 3.8*

---

6 A continuous random variable, X, has a probability density function defined by

$$f(x) = \begin{cases} \dfrac{1}{b-a} & a < x < b \\ 0 & \text{otherwise.} \end{cases}$$

Prove, using integration, that:

(a) $E(X) = \dfrac{1}{2}(a + b)$,

(d) $Var(X) = \dfrac{1}{12}(b - a)^2$. *Section 3.11*

Test yourself **ANSWERS**

1 (a)

(b) (i) $P(X > 0.5) = \dfrac{63}{64}$; **(ii)** $P(X = 1) = 0$.

2 (a) $F(r) = \begin{cases} 0 & r \leqslant 0 \\ \dfrac{1}{4}r^2 & 0 < r < 2 \\ 1 & r \geqslant 2 \end{cases}$ **(b)** $P(R \leqslant 1.5) = F(1.5) = \dfrac{9}{16}$.

3 (a) median = 1.19; **(b)** 90th percentile = 1.45; **(c)** $P(S < 1.2) = 0.512$.

4 (a) $E(X) = 1.125$; **(b) (i)** $E(4X) = 4.5$, **(ii)** $E(4X + 5) = 9.5$.

5 (a) (i) $E(Y) = 4$, **(ii)** $E(Y^2) = 16\frac{2}{3}$;

(b) $Var(Y) = \dfrac{2}{3}$ and standard deviation $(Y) = 0.816$.

CHAPTER 4

Confidence intervals

Learning objectives

After reading this chapter, you should be able to:
■ calculate a confidence interval for the mean of a normal distribution using the t-distribution.

4.1 Introduction

In S1 chapter 6 confidence intervals for the mean of a normal distribution with known standard deviation, σ, were calculated using the formula

$$x \pm z_{\frac{\alpha}{2}} \frac{\sigma}{\sqrt{n}},$$

where z is found from normal tables and n is the sample size. This formula gives a $100(1 - \alpha)\%$ confidence interval.

Later in the same chapter we saw that if σ was not known, but a large sample was available, a standard deviation calculated from the sample could be used in place of σ. In this chapter the t distribution is used to enable a confidence interval for the mean to be calculated when only a small sample from a normal distribution with unknown standard deviation is available.

The questions set in the examination and in the exercise at the end of this chapter require knowledge from S1 as well as from this chapter. It is therefore advisable to revise S1 chapter 6 before proceeding with this chapter.

4.2 Confidence interval for the mean of a normal distribution, standard deviation unknown

A 95% confidence interval for the mean of a normal distribution is given by

$$\bar{x} \pm 1.96 \frac{\sigma}{\sqrt{n}}$$

If σ is unknown it can be estimated from the sample (provided the sample is of size two or greater). However unless n is large there will be a considerable amount of uncertainty in this estimate, s. This uncertainty can be allowed for by multiplying $\frac{s}{\sqrt{n}}$ by a number larger than 1.96. The size of the number required has been calculated and tabulated in the *t*-distribution. The required value of t will depend on the amount of uncertainty. The larger the sample, the less the uncertainty. The amount of uncertainty is measured by the **degrees of freedom**. For this unit all you need to know is that an estimate of σ from a sample of size n has $n - 1$ degrees of freedom.

> This is often referred to as Student's *t*-distribution. This is tabulated in Table 5 in the Appendix and in the AQA formulae book.

> The number of degrees of freedom is usually denoted ν.

> An estimate of a population standard deviation calculated from a random sample of size n has $n - 1$ degrees of freedom.

In S1, chapter 6 a sample of four packets of baking powder weighed

193 197 212 and 184 g.

If the standard deviation is unknown we can use a calculator to find s = 11.676, $\bar{x}$ = 196.5.

The sample is of size four and so there are three degrees of freedom. For a 95% confidence interval the appropriate value of t is 3.182.

> $P = 0.975$, $\nu = 3$

Table 5 Percentage points of Student's *t*-distribution

The table gives the values of x satisfying $P(X \leqslant x) = p$, where X is a random variable having Student's *t*-distribution with ν-degrees of freedom.

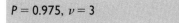

ν \ p	0.9	0.95	0.975	0.99	0.995	ν \ p	0.9	0.95	0.975	0.99	0.995
1	3.078	6.314	12.706	31.821	63.657	29	1.311	1.699	2.045	2.462	2.756
2	1.886	2.920	4.303	6.965	9.925	30	1.310	1.697	2.042	2.457	2.750
3	1.638	2.353	3.182	4.541	5.841	31	1.309	1.696	2.040	2.453	2.744
4	1.533	2.132	2.776	3.747	4.604	32	1.309	1.694	2.037	2.449	2.738
5	1.476	2.015	2.571	3.365	4.032	33	1.308	1.692	2.035	2.445	2.733

A 95% confidence interval for the mean is given by

$$196.5 \pm 3.182 \times \frac{11.676}{\sqrt{4}}$$

i.e. 196.5 ± 18.6 or 177.9 to 215.1

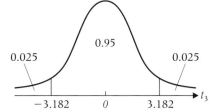

> If $\bar{x}$ is the mean of a random sample of size n from a normal distribution with mean μ, a $100(1 - \alpha)\%$ confidence interval for μ is given by
>
> $$\bar{x} \pm t_{\frac{\alpha}{2}, n-1} \frac{s}{\sqrt{n}}$$

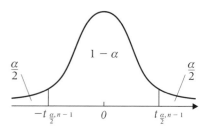

In order for the result to be valid the sample must come from a normal distribution.

Worked example 4.1

A random sample of patients who had been treated by the casualty department of a large hospital were asked to state how long they had waited before seeing a doctor. The replies, in minutes, were as follows:

    12   109   35   63   54   72   10

(a) Assuming a normal distribution calculate a 90% confidence interval for the mean stated waiting time.

(b) Comment on the hospital administrator's claim that the mean waiting time in casualty before seeing a doctor is 15 minutes.

Solution

(a) $\bar{x} = 50.71$ $s = 35.16$
sample size $= 7$ degrees of freedom $= 6$

90% confidence interval for the mean

$$50.71 \pm 1.943 \times \frac{35.16}{\sqrt{7}}$$

i.e. 50.7 ± 25.8 or 24.9 to 76.5 minutes.

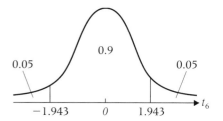

(b) As 15 lies below the confidence interval it appears that the hospital administrator underestimates the average waiting time in casualty. (However as the calculation is based on patients' recollections rather than accurate timings the data may be unreliable.)

Worked example 4.2

A machine produces plastic boxes for compact discs. The widths of the discs follow a normal distribution and the standard deviation is believed to be 0.04 mm.

A random sample of six boxes from a particular day's production is measured and found to have widths, in millimetres, of

    118.24   118.27   118.35   118.21   118.29   118.26

(a) Calculate a 95% confidence interval for the mean width of boxes produced that day:

   (i) assuming the standard deviation is 0.04 mm,

   (ii) making no assumption about the standard deviation.

(b) Give one argument in favour of using the confidence interval calculated in **(a)(i)** and one argument in favour of using the confidence interval calculated in **(a)(ii)**.

Solution

(a) **(i)** $\bar{x} = 118.270$

       95% confidence interval for mean width of boxes is

$$118.270 \pm 1.96 \times \frac{0.04}{\sqrt{6}}$$

       i.e. 118.270 ± 0.032 or 118.238 to 118.302

   (ii) $s = 0.0477$

       95% confidence interval for mean width of boxes is

$$118.270 \pm 2.571 \times \frac{0.0477}{\sqrt{6}}$$

       i.e. 118.270 ± 0.050 or 118.220 to 118.320

(b) The confidence interval calculated in **(a)(i)** uses an estimate of standard deviation based on past experience. This is likely to be more accurate than an estimate made from the current small sample. The fact that the current sample yields an estimate of standard deviation close to 0.04 means there is no reason to suppose that the standard deviation has changed.

The confidence interval calculated in **(a)(ii)** makes no assumption about the standard deviation being unchanged and is valid whether or not this is the case.

EXERCISE 4A

1 Before its annual overhaul, the mean operating time of an automatic machine was 103 s. After the annual overhaul, the following random sample of operating times (in seconds) was obtained:

    90   97   101   92   101   95   95   98   96   95

   (a) Assuming that the time taken by the machine to perform the operation is normally distributed find 95% confidence limits for the mean operating time after the overhaul.

   (b) Comment on the effectiveness of the overhaul.

2 A random sample of 12 unsalted packs of butter from a large batch had the following weights:

219 226 217 224 223 216 221 228 215 229 225 229

The weights may be assumed to be normally distributed. Find, for the mean weight of the unsalted packs of butter in the batch:

(a) an 80% confidence interval,

(b) a 90% confidence interval,

(c) a 95% confidence interval,

(d) a 99% confidence interval.

3 The resistances (in ohms) of a sample from a batch of resistors were

2314   2456   2389   2361   2360   2332   2402

Past experience suggests that the standard deviation, σ, is 35 Ω and that the distribution is normal.

(a) Calculate a 95% confidence interval for the mean resistance of the batch:
  (i) assuming $\sigma = 35$,
  (ii) making no assumption about the standard deviation.

(b) Compare the merits of the confidence intervals calculated in **(a)**. [A]

4 An automatic dispensing machine may be set to dispense predetermined quantities of liquid into bottles. Past experience suggests that the standard deviation of the amount dispensed is 2.6 ml.

An operator selected six recently filled bottles at random and measured the amounts of liquid which had been dispensed. The amounts, in millilitres, were

512   510   508   515   511   517

(a) Assuming that the sample comes from a normal distribution, calculate a 95% confidence interval for the mean if:
  (i) the standard deviation is 2.6 ml,
  (ii) no assumption is made about the standard deviation.

(b) Comment on the relative suitability in these circumstances of the two confidence intervals you have calculated. [A]

5 Stud anchors are used in the construction industry. Samples are tested by embedding them in concrete and applying a steadily increasing load until the anchor fails. A sample of six tests gave the following maximum loads in kilonewtons:

27.0   30.5   28.0   23.0   27.5   26.5

(a) Assuming a normal distribution for the maximum load calculate a 95% confidence interval for the mean.

(b) If the mean was at the lower end of the interval calculated in (a) estimate the value, k, which the maximum load would exceed with probability 0.99. Assume the standard deviation estimated is an accurate assessment of the population standard deviation.

Safety regulations require that the greatest load which may be applied under working conditions is $\dfrac{(\bar{x} - 2s)}{3}$ where $\bar{x}$ and s are calculated from a sample of six tests.

(c) Calculate this value and comment on the adequacy of this regulation in these circumstances. [A]

6 A car manufacturer introduces a new method of assembling a particular component. A sample of assembly times (minutes) taken after the new method had become established was:

27   19   68   41   17   52   35   72   38

(a) Calculate a 99% confidence interval for the mean assembly time.

(b) State any assumptions it was necessary to make in order to calculate the confidence interval in (a).

A larger random sample of 45 assembly times had a mean of 36.3 minutes with a standard deviation of 9.8 minutes.

(c) Calculate a 99% confidence for the mean assembly time.

(d) Was it necessary to make the assumptions of (b) in order to calculate the interval in (c)? Explain your answer. [A]

7 The contents of jars of honey may be assumed to be normally distributed. The contents, in grams, of a random sample of eight jars were as follows:

458   450   457   456   460   459   458   456

(a) Calculate a 95% confidence interval for the mean contents of all jars.

(b) On each jar it states 'Contents 454 grams'. Comment on this statement using the given sample and your results in (a).

(c) Given that the mean contents of all jars is 454 grams, state the probability that a 95% confidence interval calculated from the contents of a random sample of jars will not contain 454 grams. [A]

8 A firm is considering providing an unlimited supply of free bottled water for employees to drink during working hours. To estimate how much bottled water is likely to be consumed, a pilot study is undertaken. On a particular day-shift, ten employees are provided with unlimited bottled water. The amount each one consumes is monitored. The amounts, in millilitres, consumed by these ten employees are as follows:

110   0   640   790   1120   0   0   2010   830   770

(a) Assuming the data may be regarded as a random sample from a normal distribution, calculate a 95% confidence interval for the mean amount consumed on a day-shift.

(b) (i) Give a reason, based on the data collected, why the normal distribution may not provide a suitable model for the amount of free bottled water which would be consumed by employees of the firm.

    (ii) A normal distribution may provide an adequate model but cannot provide an exact model for the amount of bottled water consumed. Explain this statement giving a reason which does not depend on the data collected.

(c) Following the pilot study the firm offers free bottled water to all the 135 employees who work on the night-shift. The amounts they consume on the first night have a mean of 960 ml with a standard deviation of 240 ml.

    (i) Assuming that these data may be regarded as a random sample, calculate a 90% confidence interval for the mean amount consumed on a night-shift.

    (ii) Explain why it was not necessary to know that the data came from a normal distribution in order to calculate the confidence interval in (c)(i).

    (iii) Give two reasons why it may be unrealistic to regard the data as a random sample of the amounts that would be consumed by all employees if the scheme was introduced on all shifts on a permanent basis.

[A]

9 A health food cooperative imports a large quantity of dates and packs them into plastic bags labelled 500 grams. George, a Consumer Protection Officer, checked a random sample of the bags and found the weights, in grams, of the contents were as follows.

    497   501   486   502   492   508   489   494

(a) Assuming that weights follow a normal distribution, calculate, for the mean weight of contents of all the bags:
    (i) a 95% confidence interval,
    (ii) an 80% confidence interval.

(b) George is uncertain what conclusion to draw and so decides to start again by taking a much larger sample of the bags. He takes a random sample of 95 bags and finds they have a mean weight of 499.6 grams and a standard deviation of 9.3 grams. Use this larger sample to calculate a 90% confidence interval for the mean weight of contents of all the bags.

(c) The health food cooperative also imports raisins. George intends to take a random sample of 500 gram packets of raisins, weigh the contents and use the results to calculate an 80% and a 95% confidence interval for the mean weight, μ, of the contents of all the cooperative's packets of raisins.

    (i) Find the probability that:

        (A) the 80% confidence interval contains μ,

        (B) the 95% confidence interval contains μ but the 80% confidence interval does not.

    (ii) Instead of calculating both confidence intervals from the same sample, George now decides to calculate the 95% confidence interval from one sample and the 80% confidence interval from a second independent random sample. Find the probability that the 95% confidence interval contains μ but the 80% confidence interval does not. [A]

10 Applicants to join a police force are tested for physical fitness. Based on their performance, a physical fitness score is calculated for each applicant. Assume that the distribution of scores is normal.

(a) The scores for a random sample of ten applicants were

      55 23 44 69 22 45 54 72 34 66

Calculate a 99% confidence interval for the mean score of all applicants.

(b) The scores of a further random sample of 110 applicants had a mean of 49.5 and a standard deviation of 16.5. Use the data from this second sample to calculate:

    (i) a 95% confidence interval for the mean score of all applicants,

    (ii) an interval within which the score of approximately 95% of applicants will lie.

(c) By interpreting your results in **(b)(i)** and **(b)(ii)**, comment on the ability of the applicants to achieve a score of 25.

(d) Give two reasons why a confidence interval based on a sample of size 110 would be preferable to one based on a sample of size 10.

(e) It is suggested that a much better estimate of the mean physical fitness of all recruits could be made by combining the two samples before calculating a confidence interval. Comment on this suggestion.   [A]

Key point summary

1 An estimate of a population standard deviation   *p87*
calculated from a random sample of size n has
$n - 1$ degrees of freedom.

2 If $\bar{x}$ is the mean of a random sample of size n from   *p88*
a normal distribution with mean μ, a $100(1 - \alpha)\%$
confidence interval for μ is given by

$$\bar{x} \pm t_{\frac{\alpha}{2}, n-1} \frac{s}{\sqrt{n}}$$

Test yourself	What to review
1 A population standard deviation is estimated from a random sample of size 12. How many degrees of freedom are associated with this estimate?	*Section 4.2*
2 The weights, in grams, of a random sample of free-range eggs sold by a health food cooperative were 47 49 56 53 55 46 59 Assuming the weights follow a normal distribution, calculate a 90% confidence interval for its mean.	*Section 4.2*
3 Comment on the claim that the mean weight of the eggs in question **2** is 55.5 g.	*Section 4.2*
4 How would your answer to question **2** be affected if it was later discovered that the sample was not random?	*Section 4.2*

Test yourself   ANSWERS

4 If the sample was not random this would make the confidence interval unreliable. For example, the eggs might have been chosen because they were the largest available.

3 Although 55.5 is higher than the sample mean it lies within the confidence interval. There is therefore no convincing evidence that the mean is not 55.5 g as claimed.

2 48.5–55.8.

1 11.

CHAPTER 5

Hypothesis testing

Learning objectives

After studying this chapter, you should be able to:
- define a null and alternative hypothesis
- define the significance level of a hypothesis test
- identify a critical region
- understand whether to use a one- or a two-tailed test
- understand what is meant by a **Type I** and **Type II** error
- test a hypothesis about a population mean based on a sample from a normal distribution with known standard deviation.

> The population mean is denoted by μ.

5.1 Forming a hypothesis

One of the most important applications of statistics is to use a *sample* to test an idea, or *hypothesis*, you have regarding a population. This is one of the methods of statistical inference referred to in S1 section 6.1.

Conclusions can never be absolutely certain but the risk of your conclusion being incorrect can be quantified (measured) and can enable you to identify *statistically significant* results.

> Statistically significant results require overwhelming evidence.

In any experiment, you will have your own idea or hypothesis as to how you expect the results to turn out.

A **Null Hypothesis,** written H_0, is set up at the start of any hypothesis test. This null hypothesis is a statement which defines the population and so always contains '=' signs, never '>', '<' or '$\neq$'.

> The **null hypothesis** is only abandoned in the face of overwhelming evidence that it cannot explain the experimental results.
> Rather like in a court of law where the defendant is considered innocent until the evidence proves without doubt that he or she is guilty, the H_0 is accepted as true until test results show overwhelmingly that they cannot be explained if it was true.

An example of a Null Hypothesis which you will meet in Worked example 5.1 of this chapter is:

H_0 Population mean lifetime of bulbs, $\mu = 500$ hours.

Usually, you are hoping to show that the Null Hypothesis is **not** true and so the **Alternative Hypothesis**, written H_1, is often the hypothesis you want to establish. Worked example 5.1 has:

H_1 Population mean lifetime of bulbs, $\mu > 500$ hours.

> It often seems strange to students that they may want to show that H_0 is **not** true but, considering the examples of H_0 and H_1 given here, a manufacturer may well hope to show that bulbs have a **longer** than average lifetime.

A hypothesis test needs two hypotheses identified at the beginning: H_0 the **Null Hypothesis** and H_1 the **Alternative Hypothesis**.

H_0 states that a situation is unchanged, that a population parameter takes its usual value
H_1 states that the parameter has increased, decreased or just changed

5.2 One- and two-tailed tests

Tests which involve an H_1 with a $>$ or $<$ sign are called **one-tailed** tests because we are expecting to find just an increase or just a decrease.

Tests which involve an H_1 with a $\neq$ sign are called **two-tailed** tests as they consider any change (whether it be an increase or decrease).

For example, if data were collected on the amount of weekly pocket money given to a random selection of children aged between 12 and 14 in a rural area, and also in a city, it may be that you are interested in investigating whether children in the city are given **more** pocket money than children in rural areas. Therefore, you may set up your hypotheses as:

One-tailed tests will generally involve words such as:
better or worse,
faster or slower,
more or less,
bigger or smaller,
increase or decrease.
In Worked example 5.1,
H_1 $\mu > 500$ hours indicates a **one-tailed** test.

H_0 Population average pocket money of children is the same in the rural area and in the city, or

$$\mu \text{ (city)} = \mu \text{ (rural)}$$

H_1 Population average pocket money **greater** in city, or

$$\mu \text{ (city)} > \mu \text{ (rural)}$$

This is an example of a **one-tailed** test.

However, if you were monitoring the weight of items produced in a factory, it would be likely that **any** change, be it an increase or decrease, would be a problem and there would not necessarily be any reason to expect a change of a specific type.
In this case, typical hypotheses would be:

Two-tailed tests will generally involve words such as:
different or difference,
change,
affected.

H_0 Population mean weight is 35 g

$$\mu = 35g$$

H_1 Population mean weight **is not** 35 g

$$\mu \neq 35g$$

This is an example of a **two-tailed** test.

> A **two-tailed** test is one where H_1 involves testing for any (non-directional) change in a parameter.
> A **one-tailed** test is one where H_1 involves testing specifically for an increase or for a decrease (change in one direction only).

5.3 Testing a hypothesis about a population mean

Carrying out a hypothesis test to determine whether a population mean is significantly different from the suggested value stated in H_0 involves calculating a **test statistic** from a sample taken from the population.

As the test involves the population mean, it is the **sample mean**, $\bar{x}$, which must be evaluated. Since this test concerns a sample taken from a *normal* distribution, we also know that the sample means follow a normal distribution with mean equal to μ and with standard deviation equal to $\dfrac{\sigma}{\sqrt{n}}$.

The **test statistic** simply standardises the sample mean, $\bar{x}$, so that the result can be compared to critical z-values.

> **Test statistic** $= \dfrac{\bar{x} - \mu}{\dfrac{\sigma}{\sqrt{n}}}$

It is very important that it can be assumed or known that the sample has been selected **randomly.** If the sample were not selected randomly, then valid conclusions regarding the whole population cannot be made since the sample may only represent one part of that population.

See S1, section 5.11.

5

5.4 Critical region and significance level of test

The **critical region** is the range of values of the test statistic which is so unlikely to occur when H_0 is true, that it will lead to the conclusion that H_0 is not true. The **significance level** of a test determines what is considered the level of overwhelming evidence necessary for the decision to conclude that H_0 is not true. It is the probability of wrongly rejecting a true H_0. The smaller the significance level, the more overwhelming the evidence required. Common values used for **significance levels** are 1%, 5% or 10%.

The test introduced in this chapter is based on a sample from a normal distribution. Therefore, the **critical region** is identified by finding critical z-values from Table 4, 'Percentage

Table 4 in the AQA Formulae book.

points of the normal distribution', in exactly the same way as the z-values were found in order for confidence intervals to be constructed.

See chapter 4.

Some examples of critical regions are illustrated below.

One-tailed tests at 5% significance level

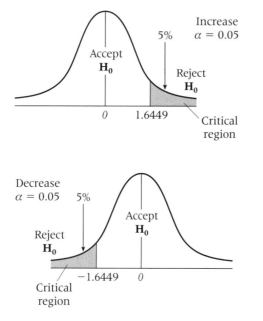

A 5% significance level is denoted $\alpha = 0.05$.

$z_\alpha = 1.6449$.

$z_\alpha = -1.6449$.

One-tailed tests at 1% significance level

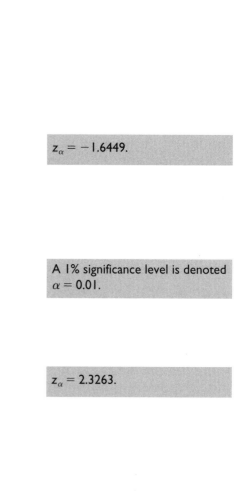

A 1% significance level is denoted $\alpha = 0.01$.

$z_\alpha = 2.3263$.

$z_\alpha = -2.3263$.

Two-tailed tests at 5% and 10% significance levels

Two critical values – change at both ends is considered.

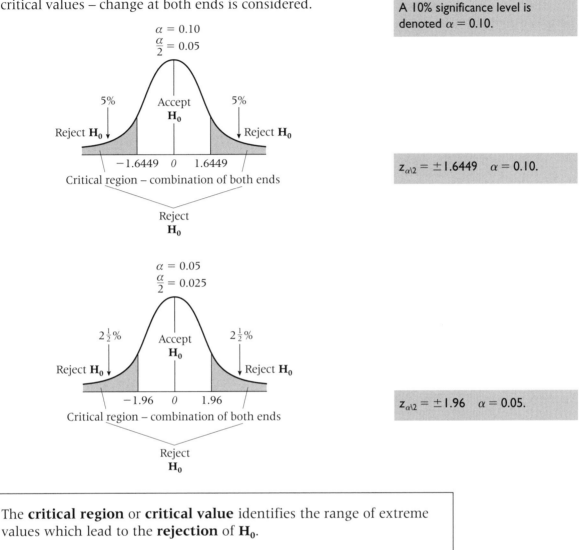

A 10% significance level is denoted $\alpha = 0.10$.

$z_{\alpha\backslash 2} = \pm 1.6449 \quad \alpha = 0.10$.

$z_{\alpha\backslash 2} = \pm 1.96 \quad \alpha = 0.05$.

The **critical region** or **critical value** identifies the range of extreme values which lead to the **rejection** of H_0.

The **significance level**, α, of a test is the probability that a test statistic lies in the extreme critical region, if H_0 is true. It determines the level of overwhelming evidence deemed necessary for the rejection of H_0.

5.5 General procedure for carrying out a hypothesis test

The general procedure for hypothesis testing is:

1 Write down H_0 and H_1 4 Identify the critical region
2 Decide which test to use   5 Calculate the test statistic
3 Decide on the significance level   6 Make your conclusion in context

Worked example 5.1

The lifetimes (hours) of Xtralong light bulbs are known to be normally distributed with a standard deviation of 90 hours.

A random sample of ten light bulbs is taken from a large batch produced in the Xtralong factory after an expensive machinery overhaul.

The lifetimes of these bulbs were measured as

      523 556 678 429 558 498 399 515 555 699 hours.

Before the overhaul the mean life was 500 hours.

Investigate, at the 5% significance level, whether the mean life of Xtralong light bulbs has increased after the overhaul.

> The bulbs may appear to have a longer mean lifetime now but this may not be statistically significant.

Solution

The important facts to note are:

We are testing whether the mean is still 500 hours or whether an increase has occurred.

This means that H_0 is $\mu = 500$ hours
              and H_1 is $\mu > 500$ hours, a **one-tailed** test.

The test is at the **5%** significance level.

The lifetimes are normally distributed with known standard deviation of 90 hours.

The hypothesis test is carried out as follows:

    H_0 $\mu = 500$ hours
    H_1 $\mu > 500$ hours $\alpha = 0.05$

From tables, the critical value is $z = 1.6449$ for this one-tailed test.

The sample mean, $\bar{x} = 541$ hours, from a sample with $n = 10$.

The population standard deviation is known, $\sigma = 90$ hours.

Therefore the test statistic

$$\frac{\bar{x} - \mu}{\frac{\sigma}{\sqrt{n}}} = \frac{541 - 500}{\frac{90}{\sqrt{10}}} = 1.44$$

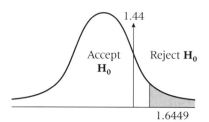

Conclusion

$1.44 < 1.6449$ for this one-tailed test, hence H_0 is accepted.

There is no significant evidence to suggest that an increase in mean lifetime has occurred since the overhaul.

> Notice that we have not **proved** that $\mu = 500$ hours but we have shown that if $\mu = 500$ hours, a sample mean of 541 is not a particularly unlikely event.

Worked example 5.2

A forestry worker decided to keep records of the first year's growth of pine seedlings. Over several years, she found that the growth followed a normal distribution with a mean of 11.5 cm and a standard deviation of 2.5 cm.

Last year, she used an experimental soil preparation for the pine seedlings and the first year's growth of a sample of eight of the seedlings was

    7   22   19   15   11   18   17   15 cm.

Investigate, at the 1% significance level, whether there has been a change in the mean growth. Assume the standard deviation has not changed.

Solution

The important facts to note are:

We are testing whether the mean is 11.5 cm or not.

This means that H_0 is $\mu = 11.5$ cm
          and H_1 is $\mu \neq 11.5$ cm a **two-tailed** test.

The test is at the **1%** significance level.

The growth is normally distributed with known standard deviation of 2.5 cm.

The hypothesis test is carried out as follows:

    H_0 $\mu = 11.5$ cm
    H_1 $\mu \neq 11.5$ cm $\alpha = 0.01$

From tables, the critical values are $z_{\frac{\alpha}{2}} = \pm 2.5758$ for this two-tailed test.

The sample mean, $\bar{x} = 15.5$, from a sample with $n = 8$.

The population standard deviation is known, $\sigma = 2.5$ cm.

Therefore the test statistic
$$\frac{\bar{x} - \mu}{\frac{\sigma}{\sqrt{n}}} = \frac{15.5 - 11.5}{\frac{2.5}{\sqrt{8}}} = 4.53$$

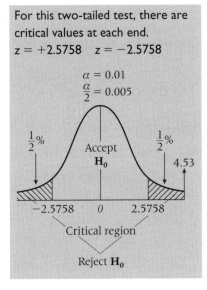

For this two-tailed test, there are critical values at each end.
$z = +2.5758$ $z = -2.5758$

$\alpha = 0.01$
$\frac{\alpha}{2} = 0.005$

$\frac{1}{2}\%$ Accept H_0 $\frac{1}{2}\%$

4.53

-2.5758 0 2.5758

Critical region

Reject H_0

Conclusion

$4.53 > 2.5758$ for this two-tailed test, hence H_0 is clearly rejected.

There is significant evidence to suggest that a change in mean growth has occurred.

It is clear from the data that the mean has **increased**. You can conclude that the mean has **changed** or that it has **increased**. Either would be accepted in an examination.

Worked example 5.3

The owner of a small vineyard has an old bottling machine which is used for filling bottles with his wine. The bottles contain a nominal 75 cl of wine.

The old machine is known to dispense volumes of wine which are normally distributed with mean 76.4 cl and a standard deviation of 0.9 cl.

The owner is concerned that his old machine is becoming unreliable and he decides to purchase a new bottling machine. The manufacturer assures the owner that the new machine will dispense volumes which are normally distributed with a standard deviation of 0.9 cl.

The owner wishes to reduce the mean volume dispensed.

A random sample of twelve 75 cl bottles are taken from a batch filled by the new machine and the volume of wine in each bottle is measured. The volumes were

75.7 76.2 75.4 75.8 75.4 76.9
76.4 75.5 76.1 76.8 76.7 76.5 cl.

Investigate, at the 5% significance level, whether the volume of wine dispensed has been reduced.

Solution

The important facts to note are:

We are testing whether the mean is still 76.4 cl or whether a decrease has occurred.

This means that $\mathbf{H_0}$ is $\mu = 76.4$ cl
and $\mathbf{H_1}$ is $\mu < 76.4$ cl a **one-tailed** test.

The test is at the **5%** significance level.

The volumes are normally distributed with known standard deviation of 0.9 cl.

The hypothesis test is carried out as follows:

$\mathbf{H_0}$ $\mu = 76.4$ cl
$\mathbf{H_1}$ $\mu < 76.4$ cl $\alpha = 0.05$

From tables, the critical value is $z = -1.6449$ for this one-tailed test.

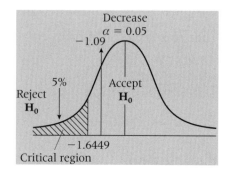

The sample mean, $\bar{x} = 76.11667$ cl, from a sample with $n = 12$.

The population standard deviation is known, $\sigma = 0.9$ cl.

Therefore the test statistic

$$\frac{\bar{x} - \mu}{\frac{\sigma}{\sqrt{n}}} = \frac{76.11667 - 76.4}{\frac{0.9}{\sqrt{12}}} = -1.09$$

Conclusion

$-1.09 > -1.6449$ for this one-tailed test, hence $\mathbf{H_0}$ is accepted.

There is no significant evidence to suggest that the mean volume has decreased.

EXERCISE 5A

1 A factory produces lengths of rope for use in boatyards. The breaking strengths of these lengths of rope follow a normal distribution with a standard deviation of 4 kg.
  The breaking strengths, in kilograms, of a random sample of 14 lengths of rope were as follows:

134	136	139	143	136	129	137
130	138	134	145	141	136	139.

  The lengths of rope are intended to have a breaking strength of 135 kg but the manufacturer claims that the mean breaking strength is in fact greater than 135 kg.
  Investigate the manufacturer's claim using a 5% level of significance.

2 Reaction times of adults in a controlled laboratory experiment are normally distributed with a standard deviation of 5 seconds. Twenty-five adults were selected at random to take part in such an experiment and the following reaction times, in seconds, were recorded:

6.5	3.4	5.6	6.9	7.1	4.9	10.9	7.8
2.4	2.8	11.3	3.7	7.8	2.4	2.8	3.7
4.9	12.0	6.5	12.8	6.9	7.4	3.1	1.9	11.5

  Investigate, using a 5% significance level, the hypothesis that the mean reaction time of adults is 7.5 s.

3 A maze is devised and after many trials, it is found that the length of time, in seconds, taken by adults to solve the maze is normally distributed with mean 7.4 and standard deviation 2.2. A group of nine children was randomly selected and asked to attempt the maze. Their times, in seconds, to completion were:

6.1	9.0	8.3	9.4	5.8	8.1	7.6	9.2	10.0

  Assuming that the times for children are also normally distributed with a standard deviation of 2.2, investigate, using the 5% significance level, whether children take longer than adults to do the maze.

4 A machine produces steel rods which are supposed to be of length 2 cm. The lengths of these rods are normally distributed with a standard deviation of 0.02 cm.
  A random sample of ten rods is taken from the production line and their lengths measured. The lengths are:

1.99	1.98	1.96	1.97	1.99	1.96	2.0	1.97	1.95	2.01 cm.

  Investigate, at the 1% significance level, whether the mean length of rods is satisfactory.

5 The resistances, in ohms, of pieces silver wire follow a normal distribution with a standard deviation of 0.02.

A random sample of nine pieces of wire are taken from a batch and their resistances were measured with results:

      1.53 1.48 1.51 1.48 1.54 1.52 1.54 1.49 1.51.

It is known that, if the wire is pure silver, the resistance should be 1.50 but, if the wire is impure, the resistance will be increased.

Investigate, at the 5% significance level, whether the batch contains pure silver wire.

6 The weight of Venus chocolate bars is normally distributed with a mean of 30 g and a standard deviation of 3.5 g.

A random sample of 20 Venus bars was taken from the production line and was found to have a mean of 32.5 g. Is there evidence, at the 1% significance level, that the mean weight has increased?

7 The weights of components produced by a certain machine are normally distributed with mean 15.4 g and standard deviation 2.3 g.

The setting on the machine is altered and, following this, a random sample of 81 components is found to have a mean weight of 15.0 g.

Does this provide evidence, at the 5% level, of a reduction in the mean weight of components produced by the machine? Assume that the standard deviation remains unaltered.

8 The ability to withstand pain is known to vary from individual to individual.

In a standard test, a tiny electric shock is applied to the finger until a tingling sensation is felt. When this test was applied to a random sample of ten adults, the times recorded, in seconds, before they experienced a tingling sensation were:

      4.2 4.5 3.9 4.4 4.1 4.5 3.7 4.8 4.2 4.2

Test, at the 5% level, the hypothesis that the mean time before an adult would experience a tingling sensation is 4.0 s. The times are known to be normally distributed with a standard deviation of 0.2 s.

5.6 Significance levels and problems to consider

You may have wondered how the **significance level** used in hypothesis testing is chosen. You have read that significance levels commonly used are 1%, 5% or 10% but no explanation has been offered about why this is so.

A common question asked by students is:

> *Why is the level of overwhelming evidence necessary to lead to rejection of H_0 commonly set at 5% ?*

The **significance level** of a hypothesis test gives the P(test statistic lies inside critical region | H_0 true). In other words, *if* H_0 is true, then, with a 5% significance level, you would expect a result as extreme as this only once in every 20 times. If the test statistic does lie in the critical region the result is statistically significant at the 5% level and we conclude that H_0 is **untrue**.

Sometimes it may be necessary to be 'more certain' of a conclusion. If a traditional trusted piece of research is to be challenged, then a 1% level of significance may be used to ensure greater confidence in rejecting H_0. If a new drug is to be used in preference to a well-known one then a 0.1% level may be necessary to ensure that no chance or fluke effects occur in research which leads to conclusions which may affect human health.

5.7 Errors

It is often quite a surprising concept for students to realise that, having correctly carried out a hypothesis test on carefully collected data and having made the relevant conclusion to accept the H_0 as true or to reject it as false, this conclusion might be right or it might be wrong.
However, you can never be absolutely certain that your conclusion is correct and has not occurred because of a *freak* result.
The significance level identifies for you the risk of a freak result leading to a wrong decision to reject H_0.
This leads many students to ask why tests so often use a **5% significance level** which has a probability of 0.05 of incorrectly rejecting H_0 when it actually is true. Why not reduce the significance level to 0.1% and then there would be a negligible risk of 0.001 of this error occurring?

The answer to this question comes from considering the **two** errors which may occur when conducting a hypothesis test. This table illustrates the problems:

		Conclusion	
		H_0 true	H_0 not true
Reality	H_0 correct	Conclusion correct	Error made **Type I**
	H_0 incorrect	Error made **Type II**	Conclusion correct

Not only can you conclude H_0 is true when really it is false but also you could conclude it is false when actually it is true.

The previous table shows that the **significance level** of a test is
P(conclusion H_0 not true | H_0 really is correct) =
P(**Type I** error made).

The other error to consider is when a test does not show a
significant result even though the H_0 actually is **not** true.
P(conclusion H_0 true | H_0 really is incorrect)
= P(**Type II** error made).

The probability of making a **Type II** error is difficult or
impossible to evaluate unless precise further information is
available about values of the population parameters. If a value
suggested in H_0 is only slightly incorrect then there may be a
very high probability of making a **Type II** error. If the value is
completely incorrect then the probability of a **Type II** error will
be very small.

> You will not be expected to
> evaluate the probability of
> making a **Type II** error.

Obviously, if you set a very low **significance level** for a test,
then the probability of making a **Type I** error will be low but
you may well have quite a high probability of making a **Type II**
error.

There is no logical reason why 5% is used, rather than 4% or 6%.
However, practical experience over a long period of time has
shown that, in most circumstances, a significance level of 5%
gives a good balance between the risks of making **Type I** and
Type II errors.

> If a low risk of wrongly rejecting
> H_0 is set, then it is unlikely that
> the test statistic will lie in the
> critical region. H_0 is unlikely to
> be rejected unless the null
> hypothesis is 'miles away' from
> reality.

This is why 5% is chosen as the 'standard' significance level for
hypothesis testing and careful consideration must be given
before changing this value.

> Errors which can occur are:
> A **Type I** error which is to reject H_0 when it is true.
> A **Type II** error which is to accept H_0 when it is not true.

Worked example 5.4

A set of times, measured to one-hundredth of a second, were
obtained from nine randomly selected subjects taking part in a
psychology experiment.

The mean of the nine sample times was found to be 9.17 s.

It is known that these times are normally distributed with a
standard deviation of 4.25 s.

The hypothesis that the mean of such times is equal to 7.50 s is
to be tested, with an alternative hypothesis that the mean is
greater than 7.50 s, using a 5% level of significance.

Explain, in the context of this situation, the meaning of:

(a) a **Type I** error,
(b) a **Type II** error.

Solution

For this example:

> H_0 $\mu = 7.50$ s
> H_1 $\mu > 7.50$ s $\alpha = 0.05$ one-tailed test.

(a) A **Type I** error is to reject H_0 and conclude that the population mean time is **greater than** 7.50 s when, in reality, the mean time for such an experiment is equal to 7.50 s.

> If H_0 is untrue it is impossible to make a **Type I** error.

The probability of this happening, if H_0 is true is $\alpha = 0.05$.

(b) A **Type II** error is to accept H_0 and conclude that the population mean time is **equal to** 7.50 s when, in reality, the mean time for such an experiment is greater than 7.50 s.

> If H_0 is true it is impossible to make a **Type II** error.

The probability of this happening will vary and can only be determined if more information is given regarding the exact alternative value that μ may take, not simply that $\mu > 7.50$ s.

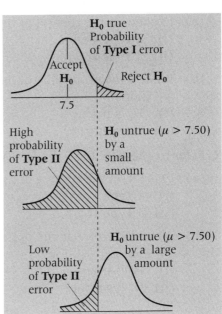

Worked example 5.5

Explain, in the context of Worked example 5.2, the meaning of:

(a) a **Type I** error,
(b) a **Type II** error.

Solution

In Worked example 5.2, we have:

> H_0 $\mu = 11.5$ cm
> H_1 $\mu \neq 11.5$ cm $\alpha = 0.01$

Hence:

(a) A **Type I** error is to reject H_0 and conclude that the mean growth of seedlings is **not equal** to 11.5 cm when, in reality, the mean growth for seedlings grown in the experimental soil preparation is equal to 11.5 cm.

The probability of this happening if H_0 is true is $\alpha = 0.01$.

(b) A **Type II** error is to accept H_0 and conclude that the mean growth of seedlings is **equal** to 11.5 cm when, in reality, the mean growth for the seedlings grown in the experimental soil preparation is not equal to 11.5 cm.

As seen in the previous example, the probability of this cannot be determined unless precise information is given regarding the alternative value taken by μ.

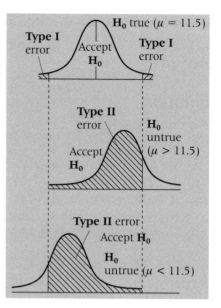

EXERCISE 5B

1 Refer to question 1 in Exercise 5A.

   (a) Explain, in the context of this question, the meaning of:
   (i) a **Type I** error,
   (ii) a **Type II** error.

   (b) What is the probability of making a **Type I** error in this question if:
   (i) H_0 is true,
   (ii) H_0 is untrue?

2 Refer to question 4 in Exercise 5A.

   (a) Explain, in the context of this question, the meaning of:
   (i) a **Type I** error,
   (ii) a **Type II** error.

   (b) What is the probability of making a **Type I** error in this question if:
   (i) H_0 is true,
   (ii) H_0 is untrue?

3 Times for glaze to set on pottery bowls follow a normal distribution with a standard deviation of three minutes. The mean time is believed to be 20 minutes.
A random sample of nine bowls is glazed with times which gave a sample mean of 18.2 minutes.

   (a) Investigate, at the 5% significance level, whether the results of this sample support the belief that the mean glaze time is 20 minutes.

   (b) Explain, in the context of this question, the meaning of a **Type I** error.

   (c) Why is it not possible to find the probability that a **Type II** error is made?

4 The lengths of car components are normally distributed with a standard deviation of 0.45 mm. Ten components are selected at random from a large batch and their lengths were found to be,

  19.3 20.5 18.1 18.3 17.6 19.0 20.1 19.2 18.6 19.4 mm.

   (a) Investigate, at the 10% significance level, the claim that the mean length of such components is 19.25 mm.

   (b) Explain, in the context of this question, the meaning of:
      (i) a **Type I** error,
      (ii) a **Type II** error.

   (c) Write down the probability of making a **Type I** error if the mean length is 19.25 mm.

5 A random sample of ten assembly workers in a large factory are trained and then asked to assemble a new design of electrical appliance. The times to assemble this appliance are known to follow a normal distribution with a standard deviation of 12 minutes. It is claimed that the new design is easier to assemble and the mean time should be less than the mean of 47 minutes currently taken to assemble the old design of appliance.
The mean time for the sample of ten workers was 39.8 minutes.

   (a) Investigate, using a 5% significance level, the claim that the new design is easier to assemble.

   (b) Explain, in the context of this question, the meaning of a **Type II** error.

Key point summary

1 A hypothesis test needs two hypotheses identified at the beginning: H_0 the **Null Hypothesis** and H_1 the **Alternative Hypothesis**.   *p96*

   H_0 and H_1 both refer to the population from which the sample is randomly taken.

2 H_0 states that a situation is unchanged, that a population parameter takes its usual value.   *p96*

   H_1 states that the parameter has increased, decreased or just changed.

   H_0 states what is to be assumed true unless overwhelming evidence proves otherwise. In the case of testing a mean, H_0 is $\mu = k$ for some suggested value of k.

3 A **two-tailed** test is one where $\mathbf{H_1}$ involves testing for any (non-directional) change in a parameter.   *p97*

A **one-tailed** test is one where $\mathbf{H_1}$ involves testing specifically for an increase or for a decrease (change in one direction only).

A **two-tailed test** results in a critical region with two areas.

A **one-tailed test** results in a critical region with one area.

4 The **critical region** or **critical value** identifies the range of extreme values which lead to the **rejection** of $\mathbf{H_0}$.   *p99*

The **critical value** is often found directly from statistical tables as in the case of testing a mean from a normally distributed population.

5 The **significance level**, α, of a test is the probability that a test statistic lies in the extreme critical region, if $\mathbf{H_0}$ is true. It determines the level of overwhelming evidence deemed necessary for the rejection of $\mathbf{H_0}$.   *p99*

The **significance level**, α, is commonly, but not exclusively, set at 1%, 5% or 10%.

6 The general procedure for hypothesis testing is: *p99*
  1 Write down $\mathbf{H_0}$ and $\mathbf{H_1}$
  2 Decide which test to use
  3 Decide on the significance level
  4 Identify the critical region
  5 Calculate the test statistic
  6 Make your conclusion in context.

7 The **test statistic** used for investigating a hypothesis regarding the mean of a normally distributed population is,   *p97*

$$\frac{\bar{x} - \mu}{\frac{\sigma}{\sqrt{n}}}$$

Where $\bar{x}$ is the mean of the randomly selected sample of size n and σ is the known population standard deviation.

If the **test statistic** lies **in** the **critical region**, or beyond the **critical value, $\mathbf{H_0}$** is rejected.

8 Errors which can occur are: *p106*

A **Type I** error which is to reject $\mathbf{H_0}$ when it is true.

A **Type II** error which is to accept $\mathbf{H_0}$ when it is not true.

The probability of making a **Type I** error is usually denoted by α.

Test yourself	**What to review**
1 Which of the following hypotheses would require a one-tailed test and which a two-tailed test?	*Section 5.2*

1 Which of the following hypotheses would require a one-tailed test and which a two-tailed test?

    (a) Amphetamines stimulate motor performance. The mean reaction time for those subjects who have taken amphetamine tablets will be faster than that for those who have not.

    (b) The mean score on a new aptitude test for a precision job is claimed to be lower than the mean of 43 found on the existing test.

    (c) Patients suffering from asthma have a higher mean health conscious index than people who do not suffer from asthma.

    (d) The mean length of rods has changed since the overhaul of a machine.

Section 5.2

2 What is the name given to the value with which a test statistic is compared in order to decide whether a null hypothesis should be rejected?

Section 5.4

3 (a) What is the name given to the agreed probability of wrongly rejecting a true null hypothesis?

    (b) Give three commonly used levels for this probability.

Section 5.4

4 A manufacturer collects data on the annual maintenance costs for a random selection of eight new welding machines. The mean cost of these eight machines is found to be £54.36. The standard deviation of such costs for welding machines is £8.74.

Stating the null and alternative hypotheses, and using a 1% significance level, test whether there is any evidence that the mean cost for maintenance of the new machines is less than the mean value of £71.90 found for the old welding machines.

Section 5.5

5 In a survey of workers who travel to work at a large factory by car, the distances, in kilometres, travelled by a random sample of ten workers were:

    14 43 17 52 22 25 68 32 26 44

In previous surveys, the mean distance was found to be 35.6 km with a standard deviation of 14.5 km.

    (a) Investigate, using a 5% significance level, whether the mean distance travelled to work has changed. Assume the standard deviation remains 14.5 km.

    (b) What is the meaning of:

        (i) a **Type I** error,

        (ii) a **Type II** error,

        in the context of this question?

    (c) Why is it important that the sample of workers is selected at random from all those factory workers who travel to work by car?

Sections 5.3, 5.5 and 5.7

5

Test yourself ANSWERS

1 **(a)** one-tail; **(b)** one-tail; **(c)** one-tail; **(d)** two-tail.

2 Critical value.

3 **(a)** Significance level; **(b)** 1%, 5%, 10%.

4 ts -5.68 cv -2.3263 mean cost is less.

5 **(a)** ts -0.284 cv ± 1.96 no change;
(b) **(i)** conclude there has been a change when in fact there has not,
(ii) conclude no change when in fact there has been a change;
(c) Conclusion unreliable if sample not random. For example, sample may have been taken only from white collar workers who may have a different mean travelling distance from manual workers.

CHAPTER 6

Further hypothesis testing for means

Learning objectives

After studying this chapter, you should be able to:

■ test a hypothesis about a population mean based on a large sample
■ test a hypothesis about a population mean based on a sample from a normal population with an unknown standard deviation.

6.1 Hypothesis test for means based on a large sample from an unspecified distribution

As discussed in S1, section 6.3, there are occasionally real life situations where we may wish to carry out a hypothesis test for a mean by examining a sample taken from a population where the standard deviation is known. However, it is much more likely that, if the mean of the population is unknown, the standard deviation will also be unknown. Provided that a large sample is available, then a sufficiently good estimate of the population standard deviation can be found from the sample. The use of a large sample also has the advantage that the sample mean is approximately normally distributed regardless of the distribution of the population.

> The definition of large is arbitrary but a sample size of $n \geqslant 30$ is usually considered 'large'.

As in chapter 5, the **sample mean**, $\bar{x}$, is evaluated and, since this test involves a **large** sample, we know that the sample mean is approximately normally distributed. The standard deviation is also evaluated and used as an estimate of σ for this test.

> It is still important to ensure that the sample is randomly selected.

The **test statistic** is $\dfrac{\bar{x} - \mu}{\dfrac{\sigma}{\sqrt{n}}}$ as before, and is compared to critical z-values.

> As the sample is large it makes very little difference whether the divisor n or $n - 1$ is used when estimating σ. However as σ is being estimated from a sample it is correct to use the divisor $n - 1$.

To carry out a hypothesis test for a mean based on a **large** sample from an **unspecified** distribution:

the **test statistic** is $\dfrac{\bar{x} - \mu}{\dfrac{\sigma}{\sqrt{n}}}$.

An estimate, s, of the standard deviation, σ, can be made from the sample, the **test statistic** is compared to **critical z-values**. These are found in AQA Table 4.

Worked example 6.1

A manufacturer claims that the mean lifetime of her batteries is 425 hours. A competitor tests a random sample of 250 of these batteries and the mean lifetime is found to be 408 hours, with a standard deviation of 68 hours.

Investigate the claim of the competitor that the batteries have a mean lifetime less than 425 hours. Use a 1% significance level.

Solution

The important facts to note are:

We are testing whether the population mean lifetime is 425 hours or whether it is less than 425 hours.

This means that $\mathbf{H_0}$ is $\mu = 425$ hours
and $\mathbf{H_1}$ is $\mu < 425$ hours a **one-tailed** test.

The test is at the **1%** significance level.

The lifetimes are from an unspecified distribution with an unknown standard deviation. However, the sample size is **large** so $\bar{x}$ is approximately normally distributed and an estimate of the unknown population standard deviation can be calculated from the sample.

$n = 250.$

The hypothesis test is carried out as follows:

$\mathbf{H_0}\ \mu = 425$ hours
$\mathbf{H_1}\ \mu < 425$ hours $\alpha = 0.01$

From tables, the critical value is $z = -2.3263$ for this one-tailed test.

The sample mean, $\bar{x} = 408$ hours, from a sample with $n = 250$.

The population standard deviation is estimated as $\sigma = 68$ hours.

The **test statistic** is $\dfrac{\bar{x} - \mu}{\dfrac{\sigma}{\sqrt{n}}} = \dfrac{408 - 425}{\dfrac{68}{\sqrt{250}}} = -3.95$

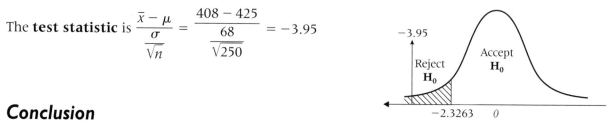

Conclusion

$-3.95 < -2.3263$ for this one-tailed test. Hence $\mathbf{H_0}$ is rejected.

There is significant evidence to suggest that the mean lifetime is less than the 425 hours claimed by the manufacturer.

Worked example 6.2

A precision machine is set to produce metal rods which have a mean length of 2 mm. A sample of 150 of these rods is randomly selected from the production of this machine. The sample mean is 1.97 mm and the standard deviation is 0.28 mm.

(a) Investigate, using the 5% significance level, the claim that the mean length of rods produced by the machine is 2 mm.

(b) How would your conclusions be affected if you later discovered that:

    (i) the sample was not random,

    (ii) the distribution was not normal?   [A]

Solution

(a) The important facts to note are:

We are testing whether or not the population mean length is 2 mm.

This is a **two-tailed** test.
The test is at the **5%** significance level.
The lengths are from an unspecified distribution with an unknown standard deviation. However, the sample size is **large**.
The hypothesis test is carried out as follows:

$$\mathbf{H_0}\ \mu = 2\ \text{mm}$$
$$\mathbf{H_1}\ \mu \neq 2\ \text{mm} \quad \alpha = 0.05$$

From tables, the critical value is $z = \pm 1.96$ for this two-tailed test.
The sample mean, $\bar{x} = 1.97$ mm, from a sample with $n = 150$.
The population standard deviation is estimated as $\sigma = 0.28$ mm.

The **test statistic** is $\dfrac{\bar{x} - \mu}{\dfrac{\sigma}{\sqrt{n}}} = \dfrac{1.97 - 2}{\dfrac{0.28}{\sqrt{150}}} = -1.31$

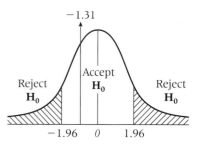

Conclusion

$-1.31 > -1.96$ for this two-tailed test. Hence $\mathbf{H_0}$ is accepted.

There is no significant evidence to suggest that a change in mean length has occurred.

We have shown that if $\mu = 2$ mm, a sample mean of 1.97 is not a particularly unlikely event.

(b) (i) If the sample was not random there can be no confidence in the conclusion. For example, these may have been the first 150 rods produced and the mean length could have changed as production continued.

    (ii) It makes no difference whether the distribution is normal or not as this is a large sample.

EXERCISE 6A

1 A factory produces lengths of cable for use in winches. These lengths are intended to have a mean breaking strength of 195 kg but the factory claims that the mean breaking strength is in fact greater than 195 kg.
In order to investigate this claim, a random sample of 80 lengths of cable is checked. The breaking strengths of these sample lengths are found to have a mean of 199.7 kg and a standard deviation of 17.4 kg. Carry out a suitable hypothesis test, using a 1% level of significance, to test the factory's claim.

2 The resistances, in ohms, of 5 mm pieces of wire for use in an electronics factory are supposed to have a mean of 1.5.
A random sample of 85 pieces of wire are taken from a large delivery and their resistances are carefully measured. Their mean resistance is 1.6 and their standard deviation is 0.9. Investigate, at the 5% significance level, whether the mean resistance is 1.5.

3 A Government department states that the mean score of Year 2 children in a new national assessment test is 78%.
A large education authority selected 90 Year 2 children at random from all those who took this test in their area and found that the mean score for these children was 72.8%, with a standard deviation of 18.5%.
Investigate, using the 5% significance level, the hypothesis that the mean score of children in this area is lower than 78%.

4 A large dairy company produces whipped spread which is packaged into 500 g plastic tubs. Sixty tubs were randomly selected from the production line and weighed. The mean weight of these tubs was found to be 504 g with a standard deviation of 17.3 g.
Investigate, using the 5% significance level, whether the mean weight of the tubs is greater than 500 g.

5 A machine produces steel rods which are supposed to be of length 30 mm. A customer suspects that the mean length of the rods is less than 30 mm. The customer selects 50 rods from a large batch and measures their lengths.
The mean length of the rods in this sample was found to be 29.1 mm with a standard deviation of 2.9 mm.
Investigate, at the 1% significance level, whether the customer's suspicions are correct.
What assumption was it necessary to make in order to carry out this test?

6 The weight of Bubbles biscuit bars is meant to have a mean of 35 g.
A random sample of 120 Bubbles bars is taken from the production line and it is found that their mean weight is 36.5 g. The standard deviation of the sample weights is 8.8 g. Is there evidence, at the 1% significance level, that the mean weight is greater than 35 g?

7 Varicoceles is a medical condition in adolescent boys which may lead to infertility. A recent edition of *The Lancet* reported a study from Italy which suggested that the presence of this condition may be detected using a surgical instrument. The instrument gives a mean reading of 7.4 for adolescent boys who do not suffer from varicoceles. A sample of 73 adolescent boys who suffered from varicoceles gave a mean reading of 6.7 and a standard deviation of 1.2.

(a) Stating your null and alternative hypotheses and using the 5% significance level, investigate whether the mean reading for all adolescent boys who suffer from varicoceles is less than 7.4.

(b) Making reference to the value of your test statistic, comment briefly on the strength of the conclusion you have drawn.

(c) State, and discuss the validity of, any assumptions you have made in **(a)** about the method of sampling and about the distribution from which the data were drawn. [A]

6.2 Hypothesis test for means based on a sample from a normal distribution with an unknown standard deviation

This is a similar situation to that encountered in section 6.1 except for the fact that the sample is **not** large. Notice also that the sample is taken from a population which is known to be normally distributed.

For large samples it was not of great importance whether the divisor n or $n - 1$ is used when calculating the standard deviation. Now, since the sample is not large, it is essential to use

$$s = \sqrt{\frac{\Sigma(x - \bar{x})^2}{n - 1}}$$ as an estimate of σ.

There is some uncertainty in using s to estimate σ when a small sample is involved. The *t*-**distribution** tables are used to find the critical values, rather than the normal distribution tables.

See Table 5 in the Appendix and in the AQA formulae book.

The test statistic is identical to that introduced in section 5.3, except for the fact that *s*, the estimated population standard deviation, is used instead of σ, the known population standard deviation.

The **test statistic** is $\dfrac{\bar{x} - \mu}{\dfrac{s}{\sqrt{n}}}$.

It is necessary to know the **degrees of freedom** before the critical value can be found. In section 4.2, you saw that an estimate of σ from a sample of size *n* has $n - 1$ **degrees of freedom**.

For example, in a one-tailed test, for an increase, at the 5% significance level using a sample of size ten, the **critical t value** is **1.833**

$(n = 10$ so the degrees of freedom, $\nu = 9$ and p $= 0.95)$

and in a two-tailed test at the 1% significance level using a sample of size 15, the **critical *t*-values** are **±2.977**

$(n = 15$ so the degrees of freedom, $\nu = 14$ and p $= 0.995)$

In the same way as this was written in section 4.2,

$$t_{\alpha,\, n-1} = t_{0.05,\, 9} = 1.833$$

and

$$t_{\frac{\alpha}{2},\, n-1} = t_{0.005,\, 14} = 2.977.$$

To carry out a hypothesis test for a mean based on a sample from a **normal** distribution with an **unknown** standard deviation:

the **test statistic** is $\dfrac{\bar{x} - \mu}{\dfrac{s}{\sqrt{n}}}$ where $s = \sqrt{\dfrac{\Sigma(x - \bar{x})^2}{n - 1}}$.

s is used to estimate σ.

The **test statistic** is compared to **critical *t*-values**. These are found in AQA Table 5. For a sample of size *n*, the degrees of freedom, $\nu = n - 1$.

Worked example 6.3

Carrots are put into cans at a large food packaging factory. The machine which puts the carrots into the cans is set so that the amount of carrots put in a can is normally distributed with a mean of 220.5 g.

The carrot packaging process is halted for two days whilst the machines have their annual overhaul. Following this overhaul, a sample of eight cans is randomly selected and the contents weighed. The weights, in grams, of carrots in each can are:

224.5 217.3 222.9 219.7 223.1 221.5 225.2 221.4

Using the 10% significance level, investigate whether there has been a change in the mean contents of these cans of carrots since the machine was overhauled.

Solution

The important facts to note are:

We are testing whether or not the population mean weight has changed from 220.5 g.

This is a **two-tailed** test.

The test is at the **10%** significance level.

The weights are from a normal distribution with an unknown standard deviation. The sample size is **not** large.

The hypothesis test is carried out as follows:

$H_0 \; \mu = 220.5$ g
$H_1 \; \mu \neq 220.5$ g $\quad \alpha = 0.10$

From tables, the critical values are $t = \pm 1.895$ for this two-tailed test.

From the sample data, $\bar{x} = 221.95$ g $\quad s = 2.577$ g $\quad n = 8$

The **test statistic** is $\dfrac{\bar{x} - \mu}{\frac{s}{\sqrt{n}}} = \dfrac{221.95 - 220.5}{\frac{2.577}{\sqrt{8}}} = 1.59$

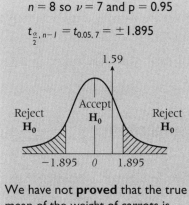

$n = 8$ so $\nu = 7$ and $p = 0.95$

$t_{\frac{\alpha}{2}, n-1} = t_{0.05, 7} = \pm 1.895$

Conclusion

$1.59 < 1.895$ for this two-tailed test. Hence H_0 is accepted.

There is no significant evidence to suggest that a change in mean weight of contents has occurred.

We have not **proved** that the true mean of the weight of carrots is 220.5 g, but there is no reason to doubt this if a sample of size eight has a mean of 221.95 g.

Worked example 6.4

Bottles of wine have nominal contents of 75 cl. The volumes of wine in bottles are normally distributed. It is suspected by a wine inspector that a particular vineyard is underfilling its wine bottles. The inspector takes a random sample of eleven bottles from the production of this vineyard and she measures the volume of wine in each bottle. The volumes, in centilitres, are found to be:

75.12   72.67   74.37   73.22   75.91   73.28
74.33   72.19   75.12   74.55   72.65

Carry out a test at the 5% significance level to investigate whether the mean volume of wine in a bottle is less than 75 cl.

Solution

The important facts to note are:

We are testing whether or not the population mean volume is less than 75 cl.

This is a **one-tailed** test.

The test is at the **5%** significance level.

The volumes are from a normal distribution with an unknown standard deviation. The sample size is **not** large.

The hypothesis test is carried out as follows:

$\mathbf{H_0}\ \mu = 75\ \text{cl}$
$\mathbf{H_1}\ \mu < 75\ \text{cl} \quad \alpha = 0.05$

From tables, the critical value is $t = -1.812$ for this one-tailed test.

From the sample data, $\bar{x} = 73.946\ \text{cl} \quad s = 1.211\ \text{cl} \quad n = 11$

The **test statistic** is $\dfrac{\bar{x} - \mu}{\dfrac{s}{\sqrt{n}}} = \dfrac{73.946 - 75}{\dfrac{1.211}{\sqrt{11}}} = -2.89$

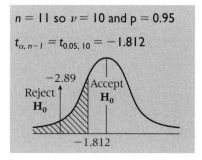

$n = 11$ so $\nu = 10$ and $p = 0.95$

$t_{\alpha,\ n-1} = t_{0.05,\ 10} = -1.812$

Conclusion

$-2.89 < -1.812$ for this one-tailed test. Hence $\mathbf{H_0}$ is rejected.

There is significant evidence to suggest that the mean volume of wine in a bottle is less than 75 cl.

Worked example 6.5

The external diameter, in centimetres, of each of a random sample of 10 piston rings manufactured on a particular machine was measured as follows:

| 9.91 | 9.89 | 10.06 | 9.98 | 10.09 |
| 9.81 | 10.01 | 9.99 | 9.87 | 10.09 |

Stating any necessary assumption, test the claim that the piston rings manufactured on this machine have a mean external diameter of 10 cm. Use a 5% significance level.

Note that the question is asking you to state an assumption.

Solution

The important facts to note are:

We are testing whether or not the population diameter is equal to 10 cm.

This is a **two-tailed** test.

The test is at the **5%** significance level.

The diameters are from an **unspecified** distribution with an unknown standard deviation. The sample size is **not** large.

We must **assume** that the diameters are **normally distributed** in order for this test to be carried out using critical *t*-values.

The hypothesis test is carried out as follows:

$\mathbf{H_0}\ \mu = 10$ cm
$\mathbf{H_1}\ \mu \neq 10$ cm $\alpha = 0.05$

From tables, the critical value is $t = \pm 2.262$ for this two-tailed test.

From the sample data, $\bar{x} = 9.97$ cm $s = 0.09695$ cm $n = 10$

The **test statistic** is $\dfrac{\bar{x} - \mu}{\dfrac{s}{\sqrt{n}}} = \dfrac{9.97 - 10}{\dfrac{0.09695}{\sqrt{10}}} = -0.978$

Conclusion

$-0.978 > -2.262$ for this two-tailed test. Hence $\mathbf{H_0}$ is accepted.

There is no significant evidence to suggest that the mean is not 10 cm.

> We are **not** told that the diameters are normally distributed.

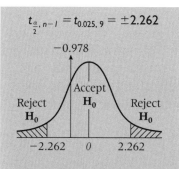

$$t_{\frac{\alpha}{2},\, n-1} = t_{0.025,\, 9} = \pm 2.262$$

> Again, we have not proved that the mean diameter is 10 cm. We just have no reason to doubt that assertion.

6

EXERCISE 6B

1 During a particular week, 13 babies were born in a maternity unit. Part of the standard procedure is to measure the length of the baby. Given below is a list of the lengths, in centimetres, of the babies born in this particular week.

    49 50 45 51 47 49 48 54 53 55 45 50 48

Assuming that this sample came from an underlying normal population, test, at the 5% significance level, the hypothesis that the population mean length is 50 cm.

2 The weights of steel ingots are known to be normally distributed. A random sample of 12 steel ingots is taken from a production line. The weights, in kilograms, of these ingots are given below.

    24.8   30.8   28.1   24.8   27.4   22.1
    24.7   27.3   27.5   27.8   23.9   23.2

Investigate the claim that the mean weight exceeds 25.0 kg using a 10% level of significance.

3 A random sample of 14 cows was selected from a large dairy herd at Brookfield Farm. Their milk yields are normally distributed. In one week the yields, in kilograms, for each cow are recorded. The results are given below.

169.6 142.0 103.3 111.6 123.4 143.5 155.1
101.7 170.7 113.2 130.9 146.1 169.3 155.5

Investigate the claim that the mean weekly milk yield for the herd is greater than 120 kg, using the 5% significance level.

4 A random sample of 15 workers from a vacuum flask assembly line was selected from a large number of such workers. Ivor Stopwatch, a work-study engineer, asked each of these workers to assemble a one litre vacuum flask at their normal working speed. The times taken, in seconds, to complete these tasks are given below.

109.2 146.2 127.9   92.0 108.5
 91.1 109.8 114.9 115.3   99.0
112.8 130.7 141.7 122.6 119.9

Assuming that this sample came from an underlying normal population, investigate the claim that the population mean assembly time is less than 2 minutes using the 5% significance level.

5 In processing grain in the brewing industry, the percentage extract recovered is measured. A brewer introduces a new source of grain and the percentage extract on 11 separate randomly selected days is as follows:

95.2 93.1 93.5 95.9 94.0 92.0
94.4 93.2 95.5 92.3 95.4

Test the hypothesis that the mean percentage extract recovered is 95.0 using the 5% significance level. What assumptions have you made in carrying out your test?

6 A car manufacturer introduces a new method of assembling a particular component. The old method had assembly times which were normally distributed with a mean of 42 minutes. The manufacturer would like the assembly time to be as short as possible, and so expects the new method to have a smaller mean. A random sample of assembly times (minutes) taken after the new method had become established was

27 39 28 41 47 42 35 32 38

Investigate the manufacturer's expectation using the 1% level of significance.

6.3 Mixed examples

In this and the previous chapter, you have been introduced to three separate sets of circumstances in which testing a mean, based on a randomly selected sample, may be carried out.

It is important to recognise which set of circumstances applies in any problem so that the correct test, either using the normal or student's *t*-distribution, is carried out.

It is also important to check whether the standard deviation of the population, σ, has been given and, if it has not, to ensure that the correct estimate for σ is used.

The following flow chart might help.

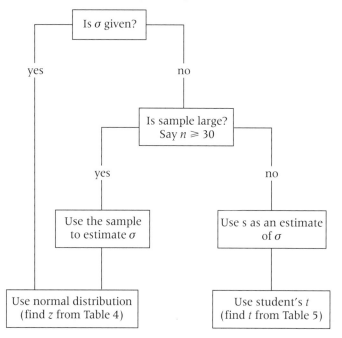

Some mixed examples are now given to illustrate how the decision making process for testing a mean works.

The mixed exercise following these examples covers different types of situations involved in testing a mean.

Worked example 6.6

A jam manufacturer produces thousands of pots of strawberry jam each week. The pots are filled by a machine such that the weights of jam delivered into the pots follow a normal distribution with mean 456 g and known standard deviation 0.7 g.

The manager of the factory feels that the mean is set too high and the machine is altered so that the mean should be reduced but the standard deviation remains unaltered.

Twelve pots of strawberry jam are taken at random from the production line following this adjustment. The amount of jam in each pot is measured with the following results in grams:

454.9 454.2 454.6 455.3 454.9 455.0
456.2 455.8 455.2 456.3 455.7 454.5

Investigate, using the 1% significance level, whether the adjustment has been satisfactory.

Solution

The important facts to note are:

We are testing whether or not the population mean weight is less than 456 g.

This is a **one-tailed** test.

The test is at the **1%** significance level.

The weights are from a **normal** distribution with a **known** standard deviation $\sigma = 0.7$ g.

The hypothesis test is carried out as follows, using the normal distribution and finding z from Table 4:

$$\mathbf{H_0} \; \mu = 456 \text{ g}$$
$$\mathbf{H_1} \; \mu < 456 \text{ g} \quad \alpha = 0.01$$

From tables, the critical value is $z = -2.3263$ for this one-tailed test.

From the sample data, $\bar{x} = 455.22$ g $\quad n = 12$

The **test statistic** is $\dfrac{\bar{x} - \mu}{\dfrac{\sigma}{\sqrt{n}}} = \dfrac{455.22 - 456}{\dfrac{0.7}{\sqrt{12}}} = -3.88$

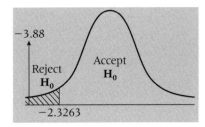

Conclusion

$-3.88 < -2.3263$ for this one-tailed test. Hence $\mathbf{H_0}$ is rejected.

There is significant evidence to suggest that the mean weight of jam delivered by the machine is less than 456 g.

Worked example 6.7

A nurse decides to record how long his journey to work takes on 55 randomly chosen days over a period of one year since the construction of new dual carriageway was started. Before this long term construction was started, his mean journey time was 44.6 minutes.

The mean time found for the 55 journeys made during the construction period is 48.5 minutes with a standard deviation of 18.4 minutes.

Investigate, at the 5% significance level, whether the construction work has led to his mean journey time being increased.

Solution

The important facts to note are:

We are testing whether or not the population mean journey time has increased.

This is a **one-tailed** test.

The test is at the **5%** significance level.

The times are from an **unspecified** distribution with an unknown standard deviation. The sample size is **large**.

$n = 55.$

The hypothesis test is carried out as follows, using the normal distribution and finding z from Table 4:

$$\mathbf{H_0} \; \mu = 44.6 \text{ minutes}$$
$$\mathbf{H_1} \; \mu > 44.6 \text{ minutes} \quad \alpha = 0.05$$

From tables, the critical value is $z = 1.6449$ for this one-tailed test.

From the sample data,
$\bar{x} = 48.5$ minutes, the estimate of σ is 18.4 minutes, $n = 55$

The **test statistic** is $\dfrac{48.5 - 44.6}{\dfrac{18.4}{\sqrt{55}}} = 1.57$

Conclusion

$1.57 < 1.6449$ for this one-tailed test. Hence $\mathbf{H_0}$ is accepted.

There is no significant evidence to suggest that the mean journey time has increased.

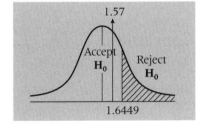

Worked example 6.8

Industrial waste dumped in rivers reduces the amount of dissolved oxygen in the water. The mean dissolved oxygen content of samples of water taken from a river at a point just above a factory suspected of illegally dumping waste is 4.9 parts per million (ppm). The dissolved oxygen contents in ppm of samples of water taken from the river at a point below the factory were,

    3.8 5.0 3.6 4.2 4.4 4.8 3.9 4.3 4.4.

(a) Stating clearly your null and alternative hypotheses, investigate at the 5% significance level whether the mean dissolved oxygen content is lower below the factory than above it. You should assume that the distribution is normal.

(b) On being shown the results the factory manager pointed out that there was a source of pollution (other than the factory) between the two points on the river which is known to reduce the dissolved oxygen content by 0.3 ppm. Does this further information affect your conclusions as to whether the factory is dumping waste or not? [A]

Solution

(a) The important facts to note are:

We are testing whether or not the population mean dissolved oxygen content is less than 4.9 ppm.

This is a **one-tailed** test.

The test is at the **5%** significance level.

The observations are from a **normal** distribution with an unknown standard deviation. The sample size is **not** large.

The hypothesis test is carried out as follows, using student's *t*-tables to find the critical value *t*:

$$\mathbf{H_0}\ \mu = 4.9$$
$$\mathbf{H_1}\ \mu < 4.9 \quad \alpha = 0.05$$

From tables, the critical value is $t = -1.860$ for this one-tailed test.

From the sample data, $\bar{x} = 4.2667 \quad s = 0.4555 \quad n = 9$

The **test statistic** is $\dfrac{\bar{x} - \mu}{\dfrac{s}{\sqrt{n}}} = \dfrac{4.2667 - 4.9}{\dfrac{0.4555}{\sqrt{9}}} = -4.17$

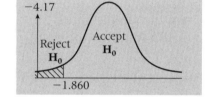

Conclusion

$-4.17 < -1.860$, hence $\mathbf{H_0}$ is rejected.

There is significant evidence to suggest that dissolved oxygen content is less below the factory than above it and hence that the factory is dumping waste.

(b) With the known source of pollution we would expect the mean dissolved oxygen content to be $4.9 - 0.3 = 4.6$ if the factory was not dumping waste.

The calculation is identical to **(a)** apart from the new hypotheses

$$\mathbf{H_0}\ \mu = 4.6$$
$$\mathbf{H_1}\ \mu < 4.6$$

The **test statistic** is $\dfrac{4.2667 - 4.6}{\dfrac{0.4555}{\sqrt{9}}} = -2.20$

The critical value is unchanged at -1.860

The evidence that the factory is dumping waste is still significant.

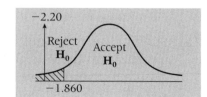

MIXED EXERCISE

1 It is known that repeated weighings of the same object on a particular chemical balance give readings which are normally distributed.

Past evidence, using experienced operators, suggests that the mean is equal to the mass of the object and that the standard deviation is 0.25 mg. A trainee operator makes seven repeated weighings of the same object, which is known to have a mass of 19.5 mg, and obtains the following readings:

19.1   19.4   19.0   18.8   19.7   19.8   19.2

Investigate using the 5% significance level, whether the trainee operator's readings are biased.

Assume the standard deviation for the trainee operator is 0.25 mg.

2 A company produces high quality chocolates which are all in the shape of circular disks.

The diameters, in millimetres, of 19 randomly selected chocolates were,

279   263   284   277   281   269   266   271   262   275
266   272   281   274   279   277   267   269   275

(a) Assuming that the diameters of these chocolates are normally distributed, investigate, at the 10% significance level, the hypothesis that their mean diameter is 275 mm.

(b) What changes would you make to your test if it was known that the standard deviation of the diameters of these chocolates was 5 mm?

3 An investigation was conducted into the dust content in the flue gases of a particular type of solid-fuel boiler. Forty boilers were used under identical fuelling and extraction conditions. Over a given period, the quantities, in grams, of dust which was deposited in traps inserted in each of the forty flues were measured. The mean quantity for this sample of forty boilers was found to be 65.7 g, with a standard deviation of 2.9 g.

Investigate, at the 1% level of significance, the hypothesis that the population mean dust deposit is 60 g.

4 A large food processing firm is considering introducing a new recipe for its ice cream. In a preliminary trial, a panel of 15 tasters were each asked to score the ice cream on a scale from 0 (awful) to 20 (excellent). Their scores were as follows:

16   15   17   6   18   15   18   7   4   16   12   14   6   17   11

The scores in a similar trial for the firm's existing ice cream were normally distributed with mean 14 and standard deviation 2.2. Assuming that the new scores are also normally distributed with standard deviation 2.2, investigate whether the mean score for the new ice cream was lower than that of the existing one. Use the 5% significance level.

6

5 As part of a research project, a random sample of 11 students sat a proposed national Physics examination and obtained the following percentage marks:

   30   44   49   50   63   38   43   96   54   40   26

Assuming such marks are normally distributed, investigate, using the 10% significance level, the hypothesis that the population mean examination mark is 40%.

6 The mean diastolic blood pressure for females is 77.4 mm. A random sample of 118 female computer operators each had their diastolic blood pressure measured. The mean diastolic blood pressure for this sample was 78.8 mm with a standard deviation of 7.1 mm. Investigate, at the 5% significance level, whether there is any evidence to suggest that female computer operators have a mean diastolic blood pressure higher than 77.4 mm.

7 The manager of a road haulage firm records the following times taken, in minutes, by a lorry to travel from the depot to a particular customer's factory.

   43   35   47   180   39   58   40   39   51

The journey time of three hours was as a result of the driver being stopped by Customs & Excise Inspectors. The manager therefore removes this value before passing the data to you, as the firm's statistician, for analysis.

(a) Use the eight remaining values to investigate, at the 5% significance level, the hypothesis that the mean time for all journeys is 40 minutes.

(b) Comment on the manager's decision to remove the value of three hours and state what assumption may have been violated if this value had been included.   [A]

8 Batteries supplied to a large institution for use in electric clocks had a mean working life of 960 days with a standard deviation of 135 days.

A sample from a new supplier had working lives of

   1040   998   894   968   890
   1280   1302   798   894   1350 hours.

Assume that the data may be regarded as a random sample from a normal distribution.

(a) Investigate whether the batteries from the new supplier have a longer working life than those from the original one. Use the 5% significance level and assume that the standard deviation of the batteries from the new supplier is also 135 days.

(b) If it cannot be assumed that the standard deviation of the batteries from the new supplier is still 135 days, explain how this would affect the test carried out in **(a)**.

(c) Carry out this new test and comment on your conclusions in **(a)** and **(c)**. [A]

9 A chamber of commerce claims that the average take-home pay of manual workers in full-time employment in its area is £140 per week. A sample of 125 such workers had mean take-home pay of £148 and standard deviation £28.

(a) Test, at the 5% significance level, the hypothesis that the mean take-home pay of all manual workers in the area is £140. Assume that the sample is random and that the distribution of take-home pay is normal. State clearly your null and alternative hypotheses.

(b) How would your conclusion be affected if you later discovered that:

    (i) the distribution of take-home pay was not normal but the sample was random,

    (ii) the sample was not random but the distribution of take-home pay was normal?

Give a brief justification for each of your answers. [A]

10 A pharmaceutical company claimed that a course of its vitamin tablets would improve examination performance. To publicise its claim, the company offered to provide the tablets free to candidates taking a particular GCSE examination. This offer was taken up by some but not all of the candidates. The average mark in the examination for all candidates who did not take the course of vitamin tablets was 42.0.

A random sample of 120 candidates from those who had taken the course of vitamin tablets gave a mean mark of 43.8 and a standard deviation of 12.8.

(a) Test, at the 5% significance level, whether the candidates who took the vitamin tablets had a mean mark greater than 42.0. State clearly your null and alternative hypotheses.

(b) Why was it not necessary to know that the examination marks were normally distributed before carrying out the test?

(c) Explain why, even if the mean mark of the sample had been much higher, the test could not prove that the course of vitamin tablets had improved examination performance. [A]

6

Key point summary

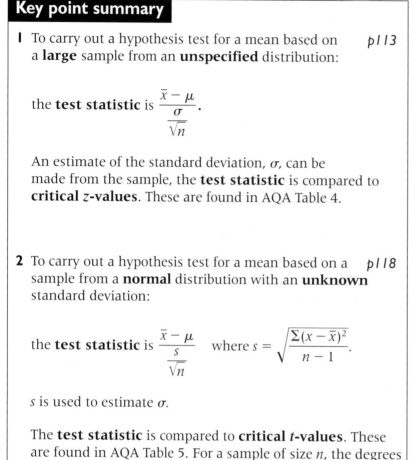

1 To carry out a hypothesis test for a mean based on   *p113*
   a **large** sample from an **unspecified** distribution:

   the **test statistic** is $\dfrac{\bar{x} - \mu}{\dfrac{\sigma}{\sqrt{n}}}$.

   An estimate of the standard deviation, σ, can be
   made from the sample, the **test statistic** is compared to
   critical z-values. These are found in AQA Table 4.

2 To carry out a hypothesis test for a mean based on a   *p118*
   sample from a **normal** distribution with an **unknown**
   standard deviation:

   the **test statistic** is $\dfrac{\bar{x} - \mu}{\dfrac{s}{\sqrt{n}}}$ where $s = \sqrt{\dfrac{\Sigma(x - \bar{x})^2}{n - 1}}$.

   s is used to estimate σ.

   The **test statistic** is compared to **critical t-values**. These
   are found in AQA Table 5. For a sample of size n, the degrees
   of freedom, $\nu = n - 1$.

Test yourself	**What to review**
1 What is the rule which is commonly used to determine whether a sample is **large**?	*Section 6.1*
2 When a population standard deviation is **unknown** and only a **small** sample is available, what is the name of the distribution which must be used in order to obtain the critical value for a test concerning the mean?	*Section 6.2*
3 State the degrees of freedom for a test concerning the mean based on a sample of size n.	*Section 6.2*
4 In each of the following cases find the critical value(s), if the hypotheses specified are tested at the 5% significance level:	*Section 6.3*

(a)
Sample mean	$\bar{x} = 32.6$
Sample size	$n = 250$
Null hypothesis	$\mathbf{H_0}\ \mu = 34$
Alternative hypothesis	$\mathbf{H_1}\ \mu \neq 34$
Population σ unknown	
Estimated standard deviation	$s = 17.6$
Population distribution unspecified	

(b)
Sample mean	$\bar{x} = 5.6$
Sample size	$n = 8$
Null hypothesis	$\mathbf{H_0}\ \mu = 5$
Alternative hypothesis	$\mathbf{H_1}\ \mu > 5$
Population σ unknown	
Estimated standard deviation	$s = 0.8$
Population has normal distribution	

(c)
Sample mean	$\bar{x} = 15.6$
Sample size	$n = 10$
Null hypothesis	$\mathbf{H_0}\ \mu = 17.5$
Alternative hypothesis	$\mathbf{H_1}\ \mu < 17.5$
Population σ unknown	
Estimated standard deviation	$s = 1.6$
Population has normal distribution	

(d)
Sample mean	$\bar{x} = 28.0$
Sample size	$n = 12$
Null hypothesis	$\mathbf{H_0}\ \mu = 30$
Alternative hypothesis	$\mathbf{H_1}\ \mu \neq 30$
Population standard deviation	$\sigma = 2.6$
Population has normal distribution	

5 Test the null hypothesis specified in question **4(b)** at the 1% significance level.

Section 6.3

6

Test yourself (continued)	**What to review**

6 Experimental components for use in aircraft engines were tested to destruction under extreme conditions. The survival times, *X* days, of ten components were as follows:

Section 6.2

207 381 411 673 534 294 597 344 418 554

(a) Investigate, at the 1% significance level, the claim that the mean survival time exceeds 400 days.

(b) What assumptions did you have to make in order to carry out the test in **(a)**?

(c) If you were told that evidence from past experience indicated that the population standard deviation of survival times is 101 days, what changes would you make to the test procedure in **(b)**?

(d) Carry out this modified test.

Test yourself **ANSWERS**

1 *n* at least 30.

2 *t*-distribution.

3 *n* − 1.

4 **(a)** ∓1.96;   **(b)** 1.895;   **(c)** −1.833;   **(d)** ∓1.96.

5 ts 2.12 cv 2.998 accept *μ* = 5.

6 **(a)** ts 0.898 cv 2.821 accept mean does not exceed 400 days;
(b) normal distribution, random sample;
(c) use *σ* = 101 instead of *s*. Critical value from normal distribution instead of *t*;
(d) ts 1.29 cv 2.326 accept mean does not exceed 400 days.

CHAPTER 7

Contingency tables

Learning objectives

After studying this chapter, you should be able to:

- analyse contingency tables using the χ^2 distribution
- recognise the conditions under which the analysis is valid
- combine classes in a contingency table to ensure the expected values are sufficiently large
- apply Yates' correction when analysing 2×2 contingency tables.

7.1 Contingency tables

A biology student observed snails on a bare limestone pavement and in a nearby limestone woodland. The colour of each snail was classified as light, medium or dark.

The following table shows the number of snails observed in each category.

	Light	Medium	Dark
Pavement	22	10	3
Woodland	8	10	12

A table, such as the one above, which shows the frequencies of two variables (colour and habitat) simultaneously is called a contingency table.

> A contingency table shows the frequencies of two (or more) variables simultaneously.

It is possible for contingency tables to show the frequencies of more than two variables. In this module you will only meet tables showing two variables.

Contingency tables are analysed to test the null hypothesis that the two variables are independent. That is, in this case, that the proportion of light snails is the same in the woodland as on the limestone pavement as are the proportions of medium snails and of dark snails. Clearly the observed proportions are not the same – for example, a much larger proportion of light snails were observed on the pavement than in the woodland. However, (as in all hypothesis testing) the hypothesis refers to the population and the test is carried out to examine whether the observed sample could reasonably have occurred by chance if the null hypothesis was true.

To carry out the test, first calculate the expected number of snails you would observe in each category if the null hypothesis were true. To do this it is helpful to extend the table to include totals and sub-totals.

	Light	Medium	Dark	Total
Pavement	22	10	3	35
Woodland	8	10	12	30
Total	30	20	15	65

row totals → 35, 30

grand total → 65

column totals ← 30, 20, 15

> The table now shows the totals for each row and for each column (the sub-totals). It also shows the total number of snails observed.

If there are the same proportion of light snails on the pavement as in the woodland then the best estimate that can be made of this proportion is the total number of light snails observed divided by the total number of snails observed. That is $\frac{30}{65}$.

Since a total of 35 snails was observed on the pavement you would expect to observe $\left(\frac{30}{65}\right) \times 35 = 16.15$ light snails on the pavement.

> The expected number refers to a long run average and so will usually not be a natural number.

Similarly you would expect $\left(\frac{30}{65}\right) \times 30$ light snails to have been observed in the woodland, $\left(\frac{20}{65}\right) \times 35$ medium snails to have been observed on the pavement, etc.

Notice that in each case the expected number in a particular cell is $\dfrac{\text{(row total)} \times \text{(column total)}}{\text{(grand total)}}$.

> The expected number in any cell of a contingency table is
> $$\frac{\text{(row total)} \times \text{(column total)}}{\text{(grand total)}}.$$

> This formula works for all contingency tables providing you are investigating the independence of two variables.

The following table shows the observed number, O, on the left of each cell and the expected number, E, on the right of each cell.

	Light	Medium	Dark	Total
Pavement	22, 16.15	10, 10.77	3, 8.08	35
Woodland	8, 13.85	10, 9.23	12, 6.92	30
Total	30	20	15	65

> It is usually sufficient to calculate the Es to two decimal places. However the more significant figures used at this stage of the calculation the better.

Note that the total of the Es is the same as the total of the Os in each row and in each column. In this case it was only necessary to derive two Es – say for the expected number of Light and Medium snails observed on the pavement – and the rest could have been deduced from the totals.

The test statistic is $X^2 = \Sigma\dfrac{(O-E)^2}{E}$. This will have a small value if the observed frequencies in each cell are close to the frequencies expected. It will have a large value if there are big differences between the frequencies observed and those expected. Hence the null hypothesis will be accepted if $X^2 = \Sigma\dfrac{(O-E)^2}{E}$ is small and rejected if it is large.

The test statistic X^2 is approximately distributed as a χ^2 distribution provided the Os are frequencies (i.e. not lengths, weights, percentages, etc.) and the Es are reasonably large (say >5). Note the contingency table must also be complete. For example, it would not be permissible to leave the dark snails out of the analysis of the contingency table above.

> $X^2 = \Sigma\dfrac{(O-E)^2}{E}$ may be approximated by the χ^2 distribution provided:
> (i) the Os are frequencies,
> (ii) the Es are reasonably large, say >5.

To obtain a critical value from the χ^2 distribution it is necessary to know the degrees of freedom. General rules exist for deriving degrees of freedom but in the case of an $m \times n$ contingency table all you need to know is that the number of degrees of freedom is $(m-1)(n-1)$.

> An $m \times n$ contingency table has $(m-1)(n-1)$ degrees of freedom.

The contingency table above is 2×3 and so there are

$(2-1)(3-1) = 2$ degrees of freedom.

Note that 2 was also the number of Es which had to be derived from the null hypothesis before the rest could be calculated from the totals. This is not a coincidence and is one way of interpreting degrees of freedom.

The analysis of the contingency table can now be completed.

	Light	Medium	Dark	Total
Pavement	22, 16.15	10, 10.77	3, 8.08	35
Woodland	8, 13.85	10, 9.23	12, 6.92	30
Total	30	20	15	65

X^2 is used as it is similar but not identical to χ^2. There is no universally recognised symbol for this statistic.

For the analysis to be valid it must be possible to allocate each snail examined to one and only one cell.

Alternatively, note that once the sub-totals are known there are only two independent frequencies. Once these are known the rest are fixed.

H_0 Colour is independent of whether a snail is found on limestone pavement or in woodland.
H_1 Colour is not independent of whether a snail is found on limestone pavement or in woodland.

$$X^2 = \Sigma\frac{(O - E)^2}{E} = \frac{(22 - 16.15)^2}{16.15} + \frac{(10 - 10.77)^2}{10.77}$$

$$+ \frac{(3 - 8.08)^2}{8.08} + \frac{(8 - 13.85)^2}{13.85}$$

$$+ \frac{(10 - 9.23)^2}{9.23} + \frac{(12 - 6.92)^2}{6.92}$$

$$= 11.6$$

You may prefer to set the calculation out in a table

O	E	$\dfrac{(O - E)^2}{E}$
22	16.15	2.1190
10	10.77	0.0551
3	8.08	3.1939
8	13.85	2.4709
10	9.23	0.0642
12	6.92	3.7292

$$X^2 = \Sigma\frac{(O - E)^2}{E} = 11.6$$

The Os are frequencies and the Es are all greater than five and so we can compare the calculated value of X^2 with a critical value from the χ^2 distribution. The degrees of freedom as calculated above are $(2 - 1)(3 - 1) = 2$.

> The test will be one-tailed since a small value of X^2 indicates good agreement between Os and Es. Only a large value of X^2 will lead to H_0 being rejected.

Table 6 Percentage points of the χ^2 distribution

The table gives the values of x satisfying $P(X \leqslant x) = p$, where X is a random variable having the χ^2 distribution with ν degrees of freedom.

ν \ p	0.005	0.01	0.025	0.05	0.1	0.9	0.95	0.975	0.99	0.995	ν \ p
1	0.00004	0.0002	0.001	0.004	0.016	2.706	3.841	5.024	6.635	7.879	1
2	0.010	0.020	0.051	0.103	0.211	4.605	5.991	7.378	9.210	10.597	2
3	0.072	0.115	0.216	0.352	0.584	6.251	7.815	9.348	11.345	12.838	3
4	0.207	0.297	0.484	0.711	1.064	7.779	9.488	11.143	13.277	14.860	4

For a 5% significance level the critical value is 5.991. Since $X^2 = 11.6$ and exceeds 5.991, H_0 is rejected and we conclude that the colour is not independent of where the snail was found. This is all that can be concluded from the hypothesis test. However, if the null hypothesis is rejected, an examination of the table will usually suggest some further interpretation which will make the result more informative. In this case the table suggests that snails found on the pavement tend to be lighter coloured than those found in the woodland.

More light snails were observed than were expected on the pavement.

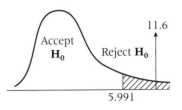

Worked example 7.1

In 1996 Prestbury School entered 45 candidates for A level statistics, while Gorton School entered 34 candidates. The following table summarises the grades obtained.

	A or B	C or D	E	N or U
Prestbury School	8	18	11	8
Gorton School	16	8	5	5

(a) Test at the 5% significance level whether the grades obtained are independent of the school.

(b) Which school has the better results? Explain your answer.

(c) Give two reasons why the school with the better results may not be the better school. [A]

Solution

(a) H_0 grades obtained are independent of school
H_1 grades obtained not independent of school

The following table shows in, each cell, the observed number on the left-hand side and the expected number (assuming the null hypothesis is true) on the right-hand side. For example the expected number obtaining A or B grades at Prestbury school is $\dfrac{45 \times 24}{79} = 13.67$

	A or B	C or D	E	N or U	Total
Prestbury School	8, 13.67	18, 14.81	11, 9.11	8, 7.41	45
Gorton School	16, 10.33	8, 11.19	5, 6.89	5, 5.59	34
Total	24	26	16	13	79

$$X^2 = \Sigma \frac{(O-E)^2}{E} = \frac{(8-13.67)^2}{13.67} + \frac{(18-14.81)^2}{14.81}$$

$$+ \frac{(11-9.11)^2}{9.11} + \frac{(8-7.41)^2}{7.41}$$

$$+ \frac{(16-10.33)^2}{10.33} + \frac{(8-11.19)^2}{11.19}$$

$$+ \frac{(5-6.89)^2}{6.89} + \frac{(5-5.59)^2}{5.59}$$

$$= 8.08$$

> The Es have been rounded to 2 dp. Despite this the calculated value of X^2 is correct to 3 sf.

The Os are frequencies and the Es are all greater than five and so we can compare the calculated value of X^2 with a critical value from the χ^2 distribution.

This is a 2×4 contingency table and so the degrees of freedom are $(2-1)(4-1) = 3$.

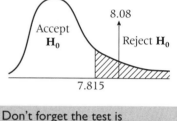

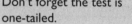

For a 5% significance level the critical value is 7.815.

H_0 is rejected and we conclude that the grades obtained are not independent of the school.

> Don't forget the test is one-tailed.

(b) From the table you can see that Prestbury school got less A or B grades than expected and more of the lower grades than expected. Gorton school got more A or B grades than expected and less of the lower grades than expected. The hypothesis test shows that this is unlikely to have occurred by chance and so you can conclude that Gorton school had the better results.

(c) A-level results are only one aspect of a school's worth and cannot on their own be used as a measure of how good a school is. The analysis takes no account of the different intakes of the two schools.

Worked example 7.2

In 1996 Ardwick School entered 130 candidates for GCSE of whom 70% gained five or more passes at grade C or above. The figures for Bramhall School were 145 candidates of whom 60% gained five or more passes at grade C or above, and for Chorlton School were 120 candidates of whom 65% gained five or more passes at grade C or above.

Form these data into a contingency table and test whether the proportion of candidates obtaining five or more passes at grade C or above is independent of the school.

Solution

A contingency table must show frequencies, not percentages and the classes must not overlap. Thus for each school it is necessary to calculate the number of candidates who gained five or more passes at grade C and above and the number of candidates who did not do so.

For Ardwick school $\dfrac{70 \times 130}{100} = 91$ candidates gained five or more passes at grade C and above and $130 - 91 = 39$ did not.

Similarly for Bramhall school $\dfrac{60 \times 145}{100} = 87$ did and $145 - 87 = 58$ did not and for Chorlton school $\dfrac{65 \times 120}{100} = 78$ did and $120 - 78 = 42$ did not.

	Ardwick	Bramhall	Chorlton
5 or more passes	91	87	78
<5 passes	39	58	42

H_0 passing five or more GCSEs at grade C is independent of school
H_1 passing five or more GCSEs at grade C is not independent of school

The following table shows in, each cell, the observed number on the left-hand side and the expected number (assuming the null hypothesis is true) on the right-hand side. For example, the expected number obtaining five or more GCSE pass grades at Ardwick school is $\dfrac{256 \times 130}{395} = 84.25$.

	Ardwick	Bramhall	Chorlton	Total
5 or more passes	91, 84.25	87, 93.97	78, 77.77	256
<5 passes	39, 45.75	58, 51.03	42, 42.23	139
Total	130	145	120	395

$$X^2 = \Sigma \frac{(O - E)^2}{E} = 3.01$$

The Os are frequencies and the Es are all greater than five and so we can compare the calculated value of X^2 with a critical value from the χ^2 distribution. This is a 2×3 contingency table and so the degrees of freedom are $(2 - 1)(3 - 1) = 2$.

For a 5% significance level the critical value is 5.991. Hence the null hypothesis is accepted and we conclude that there is no convincing evidence to show a difference in the proportions obtaining five or more GCSEs at grade C at the three schools.

> If the table showed, for each school, the total candidates and the number who gained five or more passes at grade C, it would contain the same information but it would not be a contingency table.

7

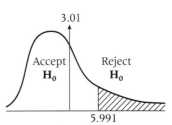

EXERCISE 7A

1 A dairy farmer kept a record of the time of delivery of each calf and the type of assistance the cow needed. Part of the data is summarised below.

	Day	Night
Unattended	42	58
Farmer assisted	63	117
Supervised by a vet	85	35

Investigate, at the 1% significance level, whether there is an association between time of day and type of birth.

2 In a survey on transport, electors from three different areas of a large city were asked whether they would prefer money to be spent on general road improvement or on improving public transport. The replies are shown in the following contingency table:

	Area		
	A	B	C
Road improvement preferred	22	46	24
Public transport preferred	78	34	36

Test, at the 1% significance level, whether the proportion favouring expenditure on general road improvement is independent of the area.

3 A statistics conference, lasting four days, was held at a university. Lunch was provided and on each day a choice of a vegetarian or a meat dish was offered for the main course. Of those taking lunch, the uptake was as follows:

	Tuesday	Wednesday	Thursday	Friday
vegetarian	17	24	21	16
meat	62	42	38	22

Test at the 5% significance level, whether the choice of dish for the main course was independent of the day of the week.

4 A survey into women's attitudes to the way in which women are portrayed in advertising was carried out for a regional television company to provide background information for a discussion programme. A questionnaire was prepared and interviewers approached women in the main shopping areas of Manchester and obtained 567 interviews.

(a) Explain why the women interviewed could not be regarded as a random sample of women living in the Manchester area.

(b) The respondents were classified by age and the following table gives the number of responses to the question 'Do you think that the way women are generally portrayed in advertising is degrading?'

	Under 35 years	35 years and over
yes definitely	85	70
yes	157	126
no	48	11
definitely not	12	3
no opinion or don't know	34	21

Use a χ^2 test, at the 5% significance level, to test whether respondents' replies are independent of age.   [A]

5 A private hospital employs a number of visiting surgeons to undertake particular operations. If complications occur during or after the operation the patient has to be transferred to the NHS hospital nearby where the required back-up facilities are available.

A hospital administrator, worried by the effects of this on costs examines the records of three surgeons. Surgeon *A* had six out of her last 47 patients transferred, surgeon *B* four out of his last 72 patients, and surgeon *C* 14 out of his last 41.

(a) Form the data into a 2 × 3 contingency table and test, at the 5% significance level, whether the proportion transferred is independent of the surgeon.

(b) The administrator decides to offer as many operations as possible to surgeon *B*. Explain why and suggest what further information you would need before deciding whether the administrator's decision was based on valid evidence.   [A]

6 The following table gives the number of candidates taking AEB Advanced Level Mathematics (Statistics) in June 1984, classified by sex and grade obtained.

	Grade obtained		
	A,B or C	D or E	O or F
Male	529	304	496
Female	398	223	271

In 1984 candidates who just failed to obtain grade E were awarded an O level. Those failing to achieve this were graded F.

(a) Use the χ^2 distribution and a 1% significance level to test whether sex and grade obtained are independent.

(b) Which sex appears to have done better? Explain your answer.

(c) The following table gives the percentage of candidates in the different grades for all the candidates taking Mathematics (Statistics) and for all the candidates taking Mathematics (Pure and Applied) in June 1984.

	Percentage of candidates		
	A, B or C	D or E	O or F
Mathematics (Statistics)	41.5	23.8	34.7
Mathematics (Pure and Applied)	29.2	26.2	44.6

(i) Explain what further information is needed before a 3×2 table can be formed from which the independence of subject and grade can be tested.

(ii) When such a table was formed the calculated value of $\Sigma \dfrac{(O - E)^2}{E}$ was 131.6. Carry out the test using a 0.5% significance level.

(iii) Discuss, briefly, whether this information indicates that it is easier to get a good grade in 'Statistics' than in 'Pure and Applied'. [A]

7 Analysis of the rate of turnover of employees by a personnel manager produced the following table showing the length of stay of 200 people who left a company for other employment.

	Length of employment (years)		
Grade	0–2	2–5	>5
Managerial	4	11	6
Skilled	32	28	21
Unskilled	25	23	50

This is a 3×3 table. The *Es* are calculated from the sub-totals in exactly the same way as in the previous examples.

Using a 1% level of significance, analyse this information and state fully your conclusions.

7.2 Small expected values

If the expected values are small then although it is still possible to calculate $\Sigma \dfrac{(O - E)^2}{E}$ it is no longer valid to obtain a critical value from the χ^2 distribution. This problem can usually be overcome by combining classes together to increase the expected values. There are three main points bear in mind when considering this procedure:

This is because if *E* is small, a small difference between *O* and *E* can lead to a relatively large value of $\dfrac{(O - E)^2}{E}$.

- It is the expected values, E, which should be reasonably large; small observed values, O, do not cause any problem
- The classes must be combined in such a way that the data remains a contingency table
- In order to be able to interpret the conclusions, classes with small Es should be combined with the most similar classes.

> A rule of thumb is that all Es should be greater than five.

> Having decided to combine classes look at the nature of the classes **not** the size of the Es in other classes.

> If it is necessary to combine classes to increase the size of the Es, the most similar classes should be combined.

Worked example 7.3

A university requires all entrants to a science course to study a non-science subject for one year. The non-science subjects available and the number of students of each sex studying them are shown in the table below.

	French	Poetry	Russian	Sculpture
Male	2	8	15	10
Female	10	17	21	37

Use a χ^2 test at the 5% significance level to test whether choice of subject is independent of gender.

Solution

H_0 choice of subject is independent of gender
H_1 choice of subject is not independent of gender

Calculating the expected values in the usual way and writing them on the right-hand side of each cell gives the following table:

	French	Poetry	Russian	Sculpture	Total
Male	2, 3.50	8, 7.29	15, 10.50	10, 13.71	35
Female	10, 8.50	17, 17.71	21, 25.50	37, 33.29	85
Total	12	25	36	47	120

> As there is only one small E and it is only a little less than five this is a borderline case. However it is safer to stick to the rule that all Es should be greater than five.

The expected value for males taking French is less than five. Choose the most similar subject to combine this with. In this case Russian is clearly the appropriate choice as it is the only other language. (We cannot just combine the French and Russian for males as this would leave three cells for males and four cells for females and there would no longer be a contingency table.)

The amended table is as follows:

	Language	Poetry	Sculpture	Total
Male	17, 14.00	8, 7.29	10, 13.71	35
Female	31, 34.00	17, 17.71	37, 33.29	85
Total	48	25	47	120

$$X^2 = \Sigma \frac{(O - E)^2}{E} = 2.42$$

There are two degrees of freedom and for a 5% significance level the critical value is 5.991. Hence the null hypothesis is accepted and we conclude that there is no convincing evidence to show a difference in the choices of males and females.

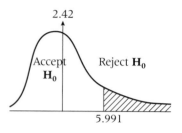

Worked example 7.4

A small supermarket chain has a branch in a city centre and also at an out-of-town shopping centre eight miles away. An investigation into the mode of transport used to visit the stores by a random sample of shoppers yielded the data below.

	Mode of transport			
	Bicycle	Public transport	Private car/taxi	Walk
City centre branch	6	20	36	8
Out-of-town branch	2	9	40	1

(a) (i) Investigate, at the 5% significance level, whether the mode of transport is independent of the branch.

(ii) Describe any differences in the popularity of the different modes of transport used to visit the two branches.

The supermarket chain now extends the investigation and includes data from all of its 14 branches. A 4×14 contingency table is formed and the statistic

$$\Sigma \frac{(O - E)^2}{E}$$

is calculated, correctly, as 56.2 with no grouping of cells being necessary.

(b) Investigate the hypothesis that the mode of transport is independent of the branch:

(i) using a 5% significance level,

(ii) using a 1% significance level.

(c) Compare and explain the conclusions you have reached in (b) (i) and (ii). [A]

Solution

(a) (i) H_0 mode of transport is independent of branch
H_1 mode of transport is not independent of branch

Calculating the expected values in the usual way and writing them on the right-hand side of each cell gives the following table.

	Mode of transport				
	Bicycle	**Public transport**	**Private car/taxi**	**Walk**	**Total**
City centre	6, 4.59	20, 16.64	36, 43.61	8, 5.16	70
Out-of-town	2, 3.41	9, 12.36	40, 32.39	1, 3.84	52
Total	8	29	76	9	122

The expected numbers going by bicycle to each branch are below five. The most similar form of transport to cycling is walking since both are self-propelled. Fortunately combining cycling and walking also eliminates the problem of the expected number walking to the out-of-town branch being less than five.

The table now becomes

	Mode of transport			
	Bicycle/ Walk	**Public transport**	**Private car/taxi**	**Total**
City centre	14, 9.75	20, 16.64	36, 43.61	70
Out-of-town	3, 7.25	9, 12.36	40, 32.39	52
Total	17	29	76	122

$$X^2 = \Sigma \frac{(O - E)^2}{E} = 9.05$$

There are two degrees of freedom and for a 5% significance level the critical value is 5.991. Hence the null hypothesis is rejected and we conclude that the method of transport is not independent of branch.

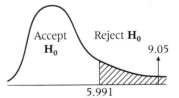

(ii) Examining the table it can be seen that at the city centre branch the observed value for bicycle/walk and for public transport exceeded the expected values whereas the observed value for private car/taxi was less than expected. It therefore appears that private cars or taxis are less likely to be used when visiting the city centre branch than when visiting the out-of-town branch.

(b) (i) The expected values for a 4×14 contingency table are calculated in exactly the same way as in any of the tables above. There are $(4 - 1)(14 - 1) = 39$ degrees of freedom. Using a 5% significance level the critical value is 54.572 which is less than 56.2 and so we would conclude that mode of transport is not independent of branch.

(ii) Using a 1% significance level the critical value is 62.428 which is greater than 56.2 and so we would accept that the mode of transport is independent of branch.

> You have been given the value of X^2 as there would not be time in an examination to calculate the Es for such a large table.

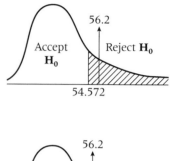

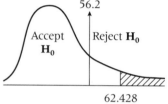

(c) The conclusions in **(b)** mean that if you are prepared to accept a 5% risk of claiming that mode of transport is not independent of branch when in fact it is independent of branch, then you can conclude that mode of transport is not independent of branch. If however you are only prepared to accept a 1% risk, there is insufficient evidence to conclude that mode of transport is not independent of branch.

EXERCISE 7B

1 A market researcher is required to interview residents of small villages, aged 18 years and over. She has been allocated a quota of 50 males and 80 females. The age and sex distribution of her interviewees are summarised below.

Age Group	Male	Female
18–29	3	5
30–39	29	21
40–49	13	40
60 and over	5	14

Investigate, using a 5% significance level, whether there is an association between the sex and age of her interviewees.

2 A manufacturer of decorating materials shows a panel of customers five different colours of paint and asks each one to identify their favourite. The choices classified by sex are shown in the table below.

	White	Pale green	Red	Dark green	Black
male	8	2	32	9	8
female	14	5	24	8	10

Test whether choice is independent of gender.

3 The number of communications received, during a particular week, by the editor of a local newspaper are shown in the table below. They have been classified by subject and by whether they were received by letter or by email.

	Politics	Football	Sport (other than football)	Miscellaneous
Letter	20	16	3	27
Email	15	8	2	10

Investigate, using a 1% significance level whether method of communication is associated with subject.

4 A survey is to be carried out among hotel guests to discover what features they regard as important when choosing an hotel. In a pilot study guests were asked to rate a number of features as important or not important. The results for four features are shown below (the number rating each feature is not the same, as not all guests were asked to rate the same features).

	Adequate lighting for reading in bed	Comfortable beds	Courteous staff	Squash courts available
important	28	34	26	4
not important	12	17	10	29

(a) Test, at the 5% significance level, whether the proportion of guests rating a feature important is independent of the feature.

(b) Comment on the relative importance of the four features.

(c) Under what circumstances would it be necessary to pool the results from more than one feature in order to carry out a valid test?

(d) If the results for 'adequate lighting for reading in bed' had to be pooled with another feature, which one would you choose and why?

(e) In the final survey 30 features were rated as unimportant, important or very important. The analysis of the resulting contingency table led to a value of 82.4 for $\sum \dfrac{(O - E)^2}{E}$ (no features were pooled). Test, at the 5% significance level, whether the rating was independent of the feature. [A]

7.3 Yates' correction for 2 × 2 contingency tables

Given the appropriate conditions $X^2 = \Sigma\dfrac{(O - E)^2}{E}$ can be approximated by a χ^2 distribution. In the case of a 2 × 2 contingency table the approximation can be improved by using $\Sigma\dfrac{(|O - E| - 0.5)^2}{E}$ instead of $\Sigma\dfrac{(O - E)^2}{E}$. This is known as Yates' correction.

> The underlying reason for this is that the Os are discrete but the χ^2 distribution is continuous. Hence this is often called Yates' continuity correction.

For a 2 × 2 table, $\Sigma\dfrac{(|O - E| - 0.5)^2}{E}$ should be calculated. This is known as Yates' correction.

> $|x|$ means the numerical value of x. Thus $|6| = 6$ and $|-3| = 3$.

Worked example 7.5

A university requires all entrants to a science course to study a non-science subject for one year. In the first year of the scheme entrants were given the choice of studying French or Russian. The number of students of each sex choosing each language is shown in the following table:

	French	Russian
Male	39	16
Female	21	14

Use a χ^2 test at the 5% significance level to test whether choice of language is independent of gender.

Solution

H$_0$ Subject chosen is independent of gender
H$_1$ Subject chosen is not independent of gender

Calculating expected values in the usual way and writing them in the right-hand side of the cell gives,

	French	Russian	Total
Male	39, 36.67	16, 18.33	55
Female	21, 23.33	14, 11.67	35
Total	60	30	90

	O	E	$O - E$	$\lvert O - E \rvert - 0.5$	$\dfrac{(\lvert O - E \rvert - 0.5)^2}{E}$
Male/French	39	36.67	2.33	1.83	0.091
Male/Russian	16	18.33	−2.33	1.83	0.183
Female/French	21	23.33	−2.33	1.83	0.144
Female/Russian	14	11.67	2.33	1.83	0.287

$$\Sigma \frac{(\lvert O - E \rvert - 0.5)^2}{E} = 0.705$$

There are $(2 - 1) \times (2 - 1) = 1$ degrees of freedom. Critical value for 5% significance level is 3.841.

Accept that choice of subject is independent of gender.

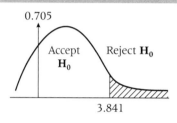

0.705

Accept $\mathbf{H_0}$ Reject $\mathbf{H_0}$

3.841

Be careful to find the modulus of $O - E$ (i.e. $\lvert O - E \rvert$) **before** subtracting 0.5.

Note: Rounding the Es to 2 d.p. may lead to small errors in the calculated value. In this case, a more accurate value is 0.707. Such small differences are of little importance.

EXERCISE 7C

1 Two groups of patients took part in an experiment in which one group received an anti-allergy drug and the other group received a placebo. The following table summarises the results:

	Drug	**Placebo**
Allergies exhibited	24	29
Allergies not exhibited	46	21

Investigate, at the 5% significance level, whether the proportion exhibiting allergies is associated with the treatment.
Is the drug effective?
Explain your answer.

2 Of 120 onion seeds of variety A planted in an allotment 28 failed to germinate. Of 45 onion seeds of variety B planted in the allotment four failed to germinate. Form the data into a 2×2 contingency table and test whether the proportion failing to germinate is associated with the variety of seed.

3 Castings made from two different moulds were tested and the results are summarised in the following table:

	Mould 1	**Mould 2**
Satisfactory	88	165
Defective	12	15

Is the proportion defective independent of mould?

7

4 As part of a research study into pattern recognition, subjects were asked to examine a picture and see if they could distinguish a word. The picture contained the word 'technology' written backwards and camouflaged by an elaborate pattern. Of 23 librarians who took part 11 succeeded in recognising the word whilst of 19 designers, 13 succeeded. Form the data into a 2 × 2 contingency table and test at the 5% significance level, using Yates' continuity correction, whether an equal proportion of librarians and designers can distinguish the word. [A]

5 The data below refer to the 1996 general election in New Zealand. They show the winning party and the percentage turnout in a sample of constituencies.

Constituency	Winning party	Percentage turnout	Constituency	Winning party	Percentage turnout
Albany	National	87.42	Hunua	National	85.50
Aoraki	Labour	87.95	Hutt South	Labour	87.85
Auckland Central	Labour	88.08	Ham	National	89.22
Banks Peninsula	National	89.71	Kaikoura	National	86.56
Bay of Plenty	National	85.23	Karapiro	National	84.23
Christchurch Cent.	Labour	83.68	Mana	Labour	87.67
Christchurch East	Labour	86.10	Mangere	Labour	79.78
Clutha-Southland	National	84.75	Maungakiekie	National	85.28
Coromandel	National	86.47	Napier	Labour	86.95
Dunedin North	Labour	88.48	Nelson	National	86.88
Dunedin South	Labour	88.87	New Lynn	Labour	85.28
Epsom	National	89.15	North Shore	National	88.34
Hamilton East	National	86.89	Otaki	Labour	88.22
Hamilton West	National	85.09	Owairaka	Labour	86.74

(a) Classify the percentage turnout as 'less than 87' or 'greater than or equal to 87'. Hence draw up a 2 × 2 contingency table suitable for testing the hypothesis that the winning party is independent of the percentage turnout.

(b) Carry out the test using a 5% significance level.

(c) Identify a feature of the constituencies in the sample which suggests that they were not randomly selected. Discuss briefly whether this is likely to affect the validity of the test carried out in **(b)**. [A]

MIXED EXERCISE

1 A well known picture of The Beatles shows them on a pedestrian crossing outside the Abbey Road recording studios. Substantial numbers of Beatles' fans visit Abbey Road and use the pedestrian crossing.

It was claimed that these fans were causing delays to rush hour traffic. As part of an investigation, people using the crossing were asked whether they were using it in the course of their normal daily lives or because of The Beatles' photograph. The answers and the times of day are summarised in the contingency table below.

	Time of day	
	Rush hour	**Out of rush hour**
Normal daily lives	34	45
Because of Beatles	4	28

(a) Using the 5% significance level, examine whether the reason for using the crossing is associated with the time of day.

(b) Interpret your result in the context of this question.

2 School inspectors in a European country classify some of the lessons they observe as unsatisfactory. Four inspectors worked as a team. Joe Stern observed 138 lessons and classified 13 as unsatisfactory, Janet Grim observed 114 lessons and classified 12 as unsatisfactory, Chris Rough observed 96 lessons and classified 28 as unsatisfactory and Ann de Sade observed 108 lessons and classified 9 as unsatisfactory.

Form the data into a contingency table. Hence investigate, at the 1% significance level, whether the proportion of lessons classified as unsatisfactory is independent of the inspector.

3 A college office stocks statistical tables and various items of stationery for sale to students.
The number of items sold by the office and the percentage of these which were statistical tables are shown in the table below. The table gives figures for each of the first five weeks of the autumn term.

Week	1	2	3	4	5
Total number of items sold	216	200	166	105	64
% of items which were statistical tables	11	16	12	17	14

(The percentages have been rounded to the nearest whole number.)

(a) Form the data above into a table suitable for testing whether the proportion of items which were statistical tables is independent of the week.

(b) Carry out this test at the 5% significance level. [A]

4 A biology student observed snails on a bare limestone pavement and in a nearby woodland. The colour of each snail was classified as light or dark. The following table shows the number of snails observed in each category.

	Light	Dark
Pavement	17	10
Woodland	3	15

(a) Use a χ^2 test at the 5% significance level to investigate whether there is an association between the colour of snails and their habitat.

(b) Describe, briefly, the nature of the association, if any, between colour and habitat.

The biology teacher complained that, since fewer than five snails were observed in the Light/Woodland cell, the test carried out in **(a)** was not valid.

(c) Comment on this complaint.

The weight, in grams, of a randomly selected snail in each category is recorded in the following table:

	Light	Dark
Pavement	14	12
Woodland	18	21

(d) The geography teacher suggested that a χ^2 test should be applied to these data. Comment on this suggestion.   [A]

5 As part of a social survey, one thousand randomly selected school leavers were sent a postal questionnaire in 1996. Completed questionnaires were returned by 712 school leavers. These 712 school leavers were asked to complete a further questionnaire in 1997. The table below shows their response to the 1997 questionnaire classified by their answers to a question on truancy in the 1996 survey.

	Persistent truant	Occasional truant	Never truant
Returned 1997 questionnaire	17	152	295
Failed to return 1997 questionnaire	23	104	121

Use the χ^2 distribution, at the 5% significance level, to test whether returning the 1997 questionnaire is independent of the answer to the 1996 question on truancy.

6 The following data are from *The British Medical Journal*. The table shows whether or not the subjects suffered from heart disease and how their snoring habits were classified by their partners.

	Never snores	Occasionally snores	Snores nearly every night	Snores every night
Heart disease	24	35	21	30
No heart disease	1355	603	192	224

(a) Use a χ^2 test, at the 5% significance level, to investigate whether frequency of snoring is related to heart disease.

(b) On the evidence above, do heart disease sufferers tend to snore more or snore less than others? Give a reason for your answer.

(c) Do these data show that snoring causes heart disease? Explain your answer briefly. [A]

7 An incurable illness, which is not life threatening, is usually treated with drugs to alleviate painful symptoms. A number of sufferers agreed to be placed at random into two groups. The members of one group would undergo a new treatment which involves major surgery and the members of the other group would continue with the standard drug treatment. Twelve months later a study of these sufferers produced information on their symptoms which is summarised in the following 2×4 contingency table:

	No change	Slight improvement	Marked improvement	Information unobtainable
New treatment	12	32	46	44
Standard treatment	36	39	12	33

(a) Test at the 5% significance level whether the outcome is independent of the treatment.

(b) Comment on the effectiveness of the new treatment in the light of your answer to (a).

(c) Further analysis of the reasons for information being unobtainable from sufferers showed that of the 44 who underwent the new treatment 19 had died, 10 had refused to cooperate and the rest were untraceable. Of the 33 who continued with the standard treatment three had died, 12 had refused to cooperate and the rest were untraceable. Form these data into a 2×3 contingency table but do not carry out any further calculations. Given that for the contingency table you have formed $\sum \frac{(O-E)^2}{E} = 10.74$, test, at the 5% significance level, whether the reason for information being unobtainable is independent of the treatment.

(d) Comment on the effectiveness of the new treatment in the light of all the information in this question. [A]

Key point summary

1 A contingency table shows the frequencies of two (or more) variables simultaneously.   *p133*

2 The expected number in any cell of a contingency table is   *p134*

$$\frac{(\text{row total}) \times (\text{column total})}{(\text{grand total})}.$$

3 $X^2 = \Sigma \dfrac{(O - E)^2}{E}$ may be approximated by the   *p135*

   χ^2 distribution provided:

   (i) the Os are frequencies,

   (ii) the Es are reasonably large, say >5.

4 An $m \times n$ contingency table has $(m - 1)(n - 1)$ degrees of freedom.   *p135*

5 If it is necessary to combine classes to increase the size of the Es the most similar classes should be combined.   *p143*

6 For a 2 $\times$2 table, $\Sigma \dfrac{(|O - E| - 0.5)^2}{E}$ should be   *p148*
   calculated. This is known as Yates' correction.

Test yourself	What to review
1 How many degrees of freedom has a 4 $\times$ 3 contingency table?	*Section 7.1*
2 When is Yates' correction applied?	*Section 7.3*
3 Find the appropriate critical value for analysing a 5 $\times$ 4 contingency table, using a 1% significance level.	*Section 7.1*
4 Why are two-sided tests not generally used when analysing contingency tables?	*Section 7.1*
5 A 2 $\times$ 4 contingency contains one cell with an expected value less than five. Why is it incorrect to combine this cell with a neighbouring cell and calculate $X^2 = \Sigma \dfrac{(O - E)^2}{E}$ for the resulting seven cells?	*Section 7.2*
6 Forty males and 50 females choose their favourite colour. A total of 24 choose red. If the results are tabulated in a contingency table, find the expected value for the female/red cell.	*Section 7.2*

1 6.

2 When a 2×2 contingency table is analysed.

3 26.2.

4 A low value of $\Sigma \dfrac{(O - E)^2}{E}$ indicates very good agreement between observed and expected values and hence is not a reason for rejecting the null hypothesis.

5 The resulting seven cells would no longer form a contingency table.

6 13.3.

Exam style practice paper

Time allowed: 1 hour 30 minutes – candidates taking the course work option will sit a slightly shorter paper (1 hour 15 minutes) of the same standard.

Answer **all** questions

---

1 The discrete random variable X has the probability distribution defined by

x	1	2	4	8	16
$P(X = x)$	0.1	0.15	0.25	0.3	0.2

   (a) Calculate the mean and variance of X.   *(3 marks)*

   (b) **(i)** Write down, in the form of a table, the probability distribution for $T = \dfrac{16}{X}$.   *(1 mark)*

        (ii) Hence calculate the value of E(T).   *(1 mark)*

   (c) A rectangle has sides of length $4(X + 1)$ and $\dfrac{4}{X}$.

        (i) Write down an expression for the area, A, of the rectangle in terms of X.   *(1 mark)*

        (ii) Hence find the value of E(A).   *(2 marks)*

2 The continuous random variable T has the following probability density function:

$$f(t) = \begin{cases} 3kt(8 - t) & 0 < t \leqslant 1 \\ k(25 - 4t) & 1 < t \leqslant 6 \\ 0 & \text{otherwise,} \end{cases}$$

where k is a constant.

   (a) Show that the cumulative distribution function of T can be written as:

$$F(t) = \begin{cases} 0 & t \leqslant 0 \\ kt^2(12 - t) & 0 < t \leqslant 1 \\ k(25t - 2t^2 - 12) & 1 < t \leqslant 6 \\ 1 & t > 6. \end{cases}$$

                                                      (5 marks)

   (b) Given that $P(T \leqslant 1) = \dfrac{1}{6}$, show that the value of the constant k is $\dfrac{1}{66}$.   *(2 marks)*

   (c) Calculate the value of the median of T.   *(4 marks)*

3 The lengths of fish of a particular species are normally distributed with a standard deviation of 4 centimetres. The mean length was 45 centimetres. There is a suspicion that, due to a change to their environment, the mean length of these fish has changed.

A random sample of 50 fish is measured and is found to have a mean length of 46.6 centimetres.

(a) Investigate, at the 1% level of significance, whether this indicates a change in the mean length of this species of fish from 45 centimetres. *(7 marks)*

(b) Explain, in context, the meaning of a Type I error. *(1 mark)*

4 Jodie has to decide which of two colleges P or Q she is going to attend in order to further her studies in Year 11.

To help her make an informed choice she decides to consider the numbers of A-level students, at each college, who achieved at least one grade A last year. These are tabulated below.

	Achieved at least one Grade A	Did not achieve at least one Grade A	Total
College P	138	160	298
College Q	127	175	302
Total	265	335	600

Use a χ^2 at the 5% level of significance to determine whether there is an association between the colleges and the proportion of A-level candidates gaining at least one grade A. *(10 marks)*

5 The marks obtained by students at Glossy High School in their mathematics examinations can be modelled by a normal distribution with mean 76.

The Mathematics department are informed that this year they have a particularly gifted group of students who should perform much better than those in previous years.

In order to investigate this claim, a random sample of 10 students was selected to take a mathematics examination, with the following results:

86   75   98   72   69   90   86   65   70   68

(a) Investigate the claim, at the 1% level of significance, that the mean mark has increased from 76. *(8 marks)*

(b) Explain, in context, the meaning of a Type II error. *(2 marks)*

6 The number of Heavy Goods Vehicles, H, passing a school each minute can be modelled by a Poisson distribution with a mean of 0.9.

   (a) Calculate $P(H = 4)$. *(2 marks)*

   (b) **(i)** Write down the distribution of X, the number of Heavy Goods Vehicles passing the school in a ten-minute period. *(1 mark)*

   (ii) Calculate $P(X \geqslant 12)$. *(2 marks)*

   (iii) Calculate the probability that at least 12 Heavy Goods Vehicles will pass the school in each of three successive ten-minute periods. *(2 marks)*

7 The heights, Y metres, of young apple trees being sold at a garden centre are normally distributed.

   The heights, in metres, of a random sample of eight young apples trees being sold at this garden centre revealed the following results:

   1.42   1.54   1.50   1.62   1.56   1.47   1.53   1.40

   (a) Calculate unbiased estimates for the mean and variance of the heights of young apple trees being sold by this garden centre. *(3 marks)*

   (b) Construct a 95% confidence interval for the mean height of young apple trees being sold by this garden centre. *(5 marks)*

   (c) In a recent advertisement, the garden centre claimed the mean height of their young apple trees to be 1.40 metres. Comment, in context, on this claim. *(2 marks)*

8 A random variable R has a probability density function defined by
$$f(r) = \begin{cases} c & a < r < b \\ 0 & \text{otherwise,} \end{cases}$$
where c is a constant.

   (a) Write down an expression for c in terms of a and b. *(1 mark)*

   (b) Given that $E(R) = \dfrac{1}{2}(a + b)$, prove, using integration, that
   $$\text{Var}(R) = \frac{1}{12}(b - a)^2. \qquad \textit{(5 marks)}$$

   (c) The error, in millimetres, made by an official measuring the distances achieved by competitors taking part in the long jump competition can be modelled by the random variable, X, with the following probability density function:
   $$f(x) = \begin{cases} k & -3 \leqslant x \leqslant 5 \\ 0 & \text{otherwise,} \end{cases}$$
   where k is a constant.

   (i) Determine the mean, μ, and standard deviation, σ, of X. *(3 marks)*

   (ii) Calculate $P\left(X < \dfrac{2 - \mu}{\sigma}\right)$. *(2 marks)*

Appendix

Table 2 Cumulative Poisson distribution function

The tabulated value is $P(X \leqslant x)$, where X has a Poisson distribution with mean λ.

x \ λ	0.1	0.2	0.3	0.4	0.5	0.6	0.7	0.8	0.9	1.0	1.2	1.4	1.6	1.8	x
0	0.9048	0.8187	0.7408	0.6703	0.6065	0.5488	0.4966	0.4493	0.4066	0.3679	0.3012	0.2466	0.2019	0.1653	0
1	0.9953	0.9825	0.9631	0.9384	0.9098	0.8781	0.8442	0.8088	0.7725	0.7358	0.6626	0.5918	0.5249	0.4628	1
2	0.9998	0.9989	0.9964	0.9921	0.9856	0.9769	0.9659	0.9526	0.9371	0.9197	0.8795	0.8335	0.7834	0.7306	2
3	1.0000	0.9999	0.9997	0.9992	0.9982	0.9966	0.9942	0.9909	0.9865	0.9810	0.9662	0.9463	0.9212	0.8913	3
4		1.0000	1.0000	0.9999	0.9998	0.9996	0.9992	0.9986	0.9977	0.9963	0.9923	0.9857	0.9763	0.9636	4
5				1.0000	1.0000	1.0000	0.9999	0.9998	0.9997	0.9994	0.9985	0.9968	0.9940	0.9896	5
6							1.0000	1.0000	1.0000	0.9999	0.9997	0.9994	0.9987	0.9974	6
7										1.0000	1.0000	0.9999	0.9997	0.9994	7
8												1.0000	1.0000	0.9999	8
9														1.0000	9

x \ λ	2.0	2.2	2.4	2.6	2.8	3.0	3.2	3.4	3.6	3.8	4.0	4.5	5.0	5.5	x
0	0.1353	0.1108	0.0907	0.0743	0.0608	0.0498	0.0408	0.0334	0.0273	0.0224	0.0183	0.0111	0.0067	0.0041	0
1	0.4060	0.3546	0.3084	0.2674	0.2311	0.1991	0.1712	0.1468	0.1257	0.1074	0.0916	0.0611	0.0404	0.0266	1
2	0.6767	0.6227	0.5697	0.5184	0.4695	0.4232	0.3799	0.3397	0.3027	0.2689	0.2381	0.1736	0.1247	0.0884	2
3	0.8571	0.8194	0.7787	0.7360	0.6919	0.6472	0.6025	0.5584	0.5152	0.4735	0.4335	0.3423	0.2650	0.2017	3
4	0.9473	0.9275	0.9041	0.8774	0.8477	0.8153	0.7806	0.7442	0.7064	0.6678	0.6288	0.5321	0.4405	0.3575	4
5	0.9834	0.9751	0.9643	0.9510	0.9349	0.9161	0.8946	0.8705	0.8441	0.8156	0.7851	0.7029	0.6160	0.5289	5
6	0.9955	0.9925	0.9884	0.9828	0.9756	0.9665	0.9554	0.9421	0.9267	0.9091	0.8893	0.8311	0.7622	0.6860	6
7	0.9989	0.9980	0.9967	0.9947	0.9919	0.9881	0.9832	0.9769	0.9692	0.9599	0.9489	0.9134	0.8666	0.8095	7
8	0.9998	0.9995	0.9991	0.9985	0.9976	0.9962	0.9943	0.9917	0.9883	0.9840	0.9786	0.9597	0.9319	0.8944	8
9	1.0000	0.9999	0.9998	0.9996	0.9993	0.9989	0.9982	0.9973	0.9960	0.9942	0.9919	0.9829	0.9682	0.9462	9
10		1.0000	1.0000	0.9999	0.9998	0.9997	0.9995	0.9992	0.9987	0.9981	0.9972	0.9933	0.9863	0.9747	10
11				1.0000	1.0000	0.9999	0.9999	0.9998	0.9996	0.9994	0.9991	0.9976	0.9945	0.9890	11
12						1.0000	1.0000	0.9999	0.9999	0.9998	0.9997	0.9992	0.9980	0.9955	12
13								1.0000	1.0000	1.0000	0.9999	0.9997	0.9993	0.9983	13
14											1.0000	0.9999	0.9998	0.9994	14
15												1.0000	0.9999	0.9998	15
16													1.0000	0.9999	16
17														1.0000	17

Table 2 Cumulative Poisson distribution function (cont.)

x \ λ	6.0	6.5	7.0	7.5	8.0	8.5	9.0	9.5	10.0	11.0	12.0	13.0	14.0	15.0	λ \ x
0	0.0025	0.0015	0.0009	0.0006	0.0003	0.0002	0.0001	0.0001	0.0000	0.0000	0.0000	0.0000	0.0000	0.0000	0
1	0.0174	0.0113	0.0073	0.0047	0.0030	0.0019	0.0012	0.0008	0.0005	0.0002	0.0001	0.0000	0.0000	0.0000	1
2	0.0620	0.0430	0.0296	0.0203	0.0138	0.0093	0.0062	0.0042	0.0028	0.0012	0.0005	0.0002	0.0001	0.0000	2
3	0.1512	0.1118	0.0818	0.0591	0.0424	0.0301	0.0212	0.0149	0.0103	0.0049	0.0023	0.0011	0.0005	0.0002	3
4	0.2851	0.2237	0.1730	0.1321	0.0996	0.0744	0.0550	0.0403	0.0293	0.0151	0.0076	0.0037	0.0018	0.0009	4
5	0.4457	0.3690	0.3007	0.2414	0.1912	0.1496	0.1157	0.0885	0.0671	0.0375	0.0203	0.0107	0.0055	0.0028	5
6	0.6063	0.5265	0.4497	0.3782	0.3134	0.2562	0.2068	0.1649	0.1301	0.0786	0.0458	0.0259	0.0142	0.0076	6
7	0.7440	0.6728	0.5987	0.5246	0.4530	0.3856	0.3239	0.2687	0.2202	0.1432	0.0895	0.0540	0.0316	0.0180	7
8	0.8472	0.7916	0.7291	0.6620	0.5925	0.5231	0.4557	0.3918	0.3328	0.2320	0.1550	0.0998	0.0621	0.0374	8
9	0.9161	0.8774	0.8305	0.7764	0.7166	0.6530	0.5874	0.5218	0.4579	0.3405	0.2424	0.1658	0.1094	0.0699	9
10	0.9574	0.9332	0.9015	0.8622	0.8159	0.7634	0.7060	0.6453	0.5830	0.4599	0.3472	0.2517	0.1757	0.1185	10
11	0.9799	0.9661	0.9467	0.9208	0.8881	0.8487	0.8030	0.7520	0.6968	0.5793	0.4616	0.3532	0.2600	0.1848	11
12	0.9912	0.9840	0.9730	0.9573	0.9362	0.9091	0.8758	0.8364	0.7916	0.6887	0.5760	0.4631	0.3585	0.2676	12
13	0.9964	0.9929	0.9872	0.9784	0.9658	0.9486	0.9261	0.8981	0.8645	0.7813	0.6815	0.5730	0.4644	0.3632	13
14	0.9986	0.9970	0.9943	0.9897	0.9827	0.9726	0.9585	0.9400	0.9165	0.8540	0.7720	0.6751	0.5704	0.4657	14
15	0.9995	0.9988	0.9976	0.9954	0.9918	0.9862	0.9780	0.9665	0.9513	0.9074	0.8444	0.7636	0.6694	0.5681	15
16	0.9998	0.9996	0.9990	0.9980	0.9963	0.9934	0.9889	0.9823	0.9730	0.9441	0.8987	0.8355	0.7559	0.6641	16
17	0.9999	0.9998	0.9996	0.9992	0.9984	0.9970	0.9947	0.9911	0.9857	0.9678	0.9370	0.8905	0.8272	0.7489	17
18	1.0000	0.9999	0.9999	0.9997	0.9993	0.9987	0.9976	0.9957	0.9928	0.9823	0.9626	0.9302	0.8826	0.8195	18
19		1.0000	1.0000	0.9999	0.9997	0.9995	0.9989	0.9980	0.9965	0.9907	0.9787	0.9573	0.9235	0.8752	19
20				1.0000	0.9999	0.9998	0.9996	0.9991	0.9984	0.9953	0.9884	0.9750	0.9521	0.9170	20
21					1.0000	0.9999	0.9998	0.9996	0.9993	0.9977	0.9939	0.9859	0.9712	0.9469	21
22						1.0000	0.9999	0.9999	0.9997	0.9990	0.9970	0.9924	0.9833	0.9673	22
23							1.0000	0.9999	0.9999	0.9995	0.9985	0.9960	0.9907	0.9805	23
24								1.0000	1.0000	0.9998	0.9993	0.9980	0.9950	0.9888	24
25										0.9999	0.9997	0.9990	0.9974	0.9938	25
26										1.0000	0.9999	0.9995	0.9987	0.9967	26
27											0.9999	0.9998	0.9994	0.9983	27
28											1.0000	0.9999	0.9997	0.9991	28
29												1.0000	0.9999	0.9996	29
30													0.9999	0.9998	30
31													1.0000	0.9999	31
32														1.0000	32

Table 3 Normal distribution function

The table gives the probability p that a normally distributed random variable Z, with mean $= 0$ and variance $= 1$, is less than or equal to z.

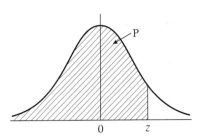

z	0.00	0.01	0.02	0.03	0.04	0.05	0.06	0.07	0.08	0.09	z
0.0	0.50000	0.50399	0.50798	0.51197	0.51595	0.51994	0.52392	0.52790	0.53188	0.53586	0.0
0.1	0.53983	0.54380	0.54776	0.55172	0.55567	0.55962	0.56356	0.56749	0.57142	0.57535	0.1
0.2	0.57926	0.58317	0.58706	0.59095	0.59483	0.59871	0.60257	0.60642	0.61026	0.61409	0.2
0.3	0.61791	0.62172	0.62552	0.62930	0.63307	0.63683	0.64058	0.64431	0.64803	0.65173	0.3
0.4	0.65542	0.65910	0.66276	0.66640	0.67003	0.67364	0.67724	0.68082	0.68439	0.68793	0.4
0.5	0.69146	0.69497	0.69847	0.70194	0.70540	0.70884	0.71226	0.71566	0.71904	0.72240	0.5
0.6	0.72575	0.72907	0.73237	0.73565	0.73891	0.74215	0.74537	0.74857	0.75175	0.75490	0.6
0.7	0.75804	0.76115	0.76424	0.76730	0.77035	0.77337	0.77637	0.77935	0.78230	0.78524	0.7
0.8	0.78814	0.79103	0.79389	0.79673	0.79955	0.80234	0.80511	0.80785	0.81057	0.81327	0.8
0.9	0.81594	0.81859	0.82121	0.82381	0.82639	0.82894	0.83147	0.83398	0.83646	0.83891	0.9
1.0	0.84134	0.84375	0.84614	0.84849	0.85083	0.85314	0.85543	0.85769	0.85993	0.86214	1.0
1.1	0.86433	0.86650	0.86864	0.87076	0.87286	0.87493	0.87698	0.87900	0.88100	0.88298	1.1
1.2	0.88493	0.88686	0.88877	0.89065	0.89251	0.89435	0.89617	0.89796	0.89973	0.90147	1.2
1.3	0.90320	0.90490	0.90658	0.90824	0.90988	0.91149	0.91309	0.91466	0.91621	0.91774	1.3
1.4	0.91924	0.92073	0.92220	0.92364	0.92507	0.92647	0.92785	0.92922	0.93056	0.93189	1.4
1.5	0.93319	0.93448	0.93574	0.93699	0.93822	0.93943	0.94062	0.94179	0.94295	0.94408	1.5
1.6	0.94520	0.94630	0.94738	0.94845	0.94950	0.95053	0.95154	0.95254	0.95352	0.95449	1.6
1.7	0.95543	0.95637	0.95728	0.95818	0.95907	0.95994	0.96080	0.96164	0.96246	0.96327	1.7
1.8	0.96407	0.96485	0.96562	0.96638	0.96712	0.96784	0.96856	0.96926	0.96995	0.97062	1.8
1.9	0.97128	0.97193	0.97257	0.97320	0.97381	0.97441	0.97500	0.97558	0.97615	0.97670	1.9
2.0	0.97725	0.97778	0.97831	0.97882	0.97932	0.97982	0.98030	0.98077	0.98124	0.98169	2.0
2.1	0.98214	0.98257	0.98300	0.98341	0.98382	0.98422	0.98461	0.98500	0.98537	0.98574	2.1
2.2	0.98610	0.98645	0.98679	0.98679	0.98713	0.98745	0.98778	0.98809	0.98840	0.98899	2.2
2.3	0.98928	0.98956	0.98983	0.99010	0.99036	0.99061	0.99086	0.99111	0.99134	0.99158	2.3
2.4	0.99180	0.99202	0.99224	0.99245	0.99266	0.99286	0.99305	0.99324	0.99343	0.99361	2.4
2.5	0.99379	0.99396	0.99413	0.99430	0.99446	0.99461	0.99477	0.99492	0.99506	0.99520	2.5
2.6	0.99534	0.99547	0.99560	0.99573	0.99585	0.99598	0.99609	0.99621	0.99632	0.99643	2.6
2.7	0.99653	0.99664	0.99674	0.99683	0.99693	0.99702	0.99711	0.99720	0.99728	0.99736	2.7
2.8	0.99744	0.99752	0.99760	0.99767	0.99774	0.99781	0.99788	0.99795	0.99801	0.99807	2.8
2.9	0.99813	0.99819	0.99825	0.99831	0.99836	0.99841	0.99846	0.99851	0.99856	0.99861	2.9
3.0	0.99865	0.99869	0.99874	0.99878	0.99882	0.99886	0.99889	0.99893	0.99896	0.99900	3.0
3.1	0.99903	0.99906	0.99910	0.99913	0.99916	0.99918	0.99921	0.99924	0.99926	0.99929	3.1
3.2	0.99931	0.99934	0.99936	0.99938	0.99940	0.99942	0.99944	0.99946	0.99948	0.99950	3.2
3.3	0.99952	0.99953	0.99955	0.99957	0.99958	0.99960	0.99961	0.99962	0.99964	0.99965	3.3
3.4	0.99966	0.99968	0.99969	0.99970	0.99971	0.99972	0.99973	0.99974	0.99975	0.99976	3.4
3.5	0.99977	0.99978	0.99978	0.99979	0.99980	0.99981	0.99981	0.99982	0.99983	0.99983	3.5
3.6	0.99984	0.99985	0.99985	0.99986	0.99986	0.99987	0.99987	0.99988	0.99988	0.99989	3.6
3.7	0.99989	0.99990	0.99990	0.99990	0.99991	0.99991	0.99992	0.99992	0.99992	0.99992	3.7
3.8	0.99993	0.99993	0.99993	0.99994	0.99994	0.99994	0.99994	0.99995	0.99995	0.99995	3.8
3.9	0.99995	0.99995	0.99996	0.99996	0.99996	0.99996	0.99996	0.99996	0.99997	0.99997	3.9

Table 4 Percentage points of the normal distribution

The table gives the values of z satisfying $P(Z \leq z) = p$, where Z is the normally distributed random variable with mean = 0 and variance = 1.

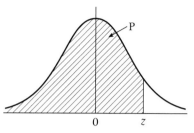

p	0.00	0.01	0.02	0.03	0.04	0.05	0.06	0.07	0.08	0.09	p
0.5	0.0000	0.0251	0.0502	0.0753	0.1004	0.1257	0.1510	0.1764	0.2019	0.2275	0.5
0.6	0.2533	0.2793	0.3055	0.3319	0.3585	0.3853	0.4125	0.4399	0.4677	0.4958	0.6
0.7	0.5244	0.5534	0.5828	0.6128	0.6433	0.6745	0.7063	0.7388	0.7722	0.8064	0.7
0.8	0.8416	0.8779	0.9154	0.9542	0.9945	1.0364	1.0803	1.1264	1.1750	1.2265	0.8
0.9	1.2816	1.3408	1.4051	1.4758	1.5548	1.6449	1.7507	1.8808	2.0537	2.3263	0.9

p	0.000	0.001	0.002	0.003	0.004	0.005	0.006	0.007	0.008	0.009	p
0.95	1.6449	1.6546	1.6646	1.6747	1.6849	1.6954	1.7060	1.7169	1.7279	1.7392	0.95
0.96	1.7507	1.7624	1.7744	1.7866	1.7991	1.8119	1.8250	1.8384	1.8522	1.8663	0.96
0.97	1.8808	1.8957	1.9110	1.9268	1.9431	1.9600	1.9774	1.9954	2.0141	2.0335	0.97
0.98	2.0537	2.0749	2.0969	2.1201	2.1444	2.1701	2.1973	2.2262	2.2571	2.2904	0.98
0.99	2.3263	2.3656	2.4089	2.4573	2.5121	2.5758	2.6521	2.7478	2.8782	3.0902	0.99

Table 5 Percentage points of Student's *t*-distribution

The table gives the values of x satisfying $P(X \leqslant x) = p$,
where X is a random variable having Student's *t*-distribution
with ν-degrees of freedom.

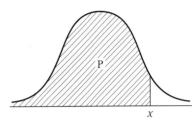

ν \ p	0.9	0.95	0.975	0.99	0.995
1	3.078	6.314	12.706	31.821	63.657
2	1.886	2.920	4.303	6.965	9.925
3	1.638	2.353	3.182	4.541	5.841
4	1.533	2.132	2.776	3.747	4.604
5	1.476	2.015	2.571	3.365	4.032
6	1.440	1.943	2.447	3.143	3.707
7	1.415	1.895	2.365	2.998	3.499
8	1.397	1.860	2.306	2.896	3.355
9	1.383	1.833	2.262	2.821	3.250
10	1.372	1.812	2.228	2.764	3.169
11	1.363	1.796	2.201	2.718	3.106
12	1.356	1.782	2.179	2.681	3.055
13	1.350	1.771	2.160	2.650	3.012
14	1.345	1.761	2.145	2.624	2.977
15	1.341	1.753	2.131	2.602	2.947
16	1.337	1.746	2.121	2.583	2.921
17	1.333	1.740	2.110	2.567	2.898
18	1.330	1.734	2.101	2.552	2.878
19	1.328	1.729	2.093	2.539	2.861
20	1.325	1.725	2.086	2.528	2.845
21	1.323	1.721	2.080	2.518	2.831
22	1.321	1.717	2.074	2.508	2.819
23	1.319	1.714	2.069	2.500	2.807
24	1.318	1.711	2.064	2.492	2.797
25	1.316	1.708	2.060	2.485	2.787
26	1.315	1.706	2.056	2.479	2.779
27	1.314	1.703	2.052	2.473	2.771
28	1.313	1.701	2.048	2.467	2.763

ν \ p	0.9	0.95	0.975	0.99	0.995
29	1.311	1.699	2.045	2.462	2.756
30	1.310	1.697	2.042	2.457	2.750
31	1.309	1.696	2.040	2.453	2.744
32	1.309	1.694	2.037	2.449	2.738
33	1.308	1.692	2.035	2.445	2.733
34	1.307	1.691	2.032	2.441	2.728
35	1.306	1.690	2.030	2.438	2.724
36	1.306	1.688	2.028	2.434	2.719
37	1.305	1.687	2.026	2.431	2.715
38	1.304	1.686	2.024	2.429	2.712
39	1.304	1.685	2.023	2.426	2.708
40	1.303	1.684	2.021	2.423	2.704
45	1.301	1.679	2.014	2.412	2.690
50	1.299	1.676	2.009	2.403	2.678
55	1.297	1.673	2.004	2.396	2.668
60	1.296	1.671	2.000	2.390	2.660
65	1.295	1.669	1.997	2.385	2.654
70	1.294	1.667	1.994	2.381	2.648
75	1.293	1.665	1.992	2.377	2.643
80	1.292	1.664	1.990	2.374	2.639
85	1.292	1.663	1.998	2.371	2.635
90	1.291	1.662	1.987	2.368	2.632
95	1.291	1.661	1.985	2.366	2.629
100	1.290	1.660	1.984	2.364	2.626
125	1.288	1.657	1.979	2.357	2.616
150	1.287	1.655	1.976	2.351	2.609
200	1.286	1.653	1.972	2.345	2.601
∞	1.282	1.645	1.960	2.326	2.576

Table 6 Percentage points of the χ^2 distribution

The table gives the values of x satisfying $P(X \leq x) = p$, where X is a random variable having the χ^2 distribution with v-degrees of freedom.

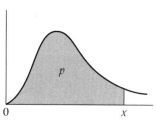

v \ p	0.005	0.01	0.025	0.05	0.1	0.9	0.95	0.975	0.99	0.995	v
1	0.00004	0.0002	0.001	0.004	0.016	2.706	3.841	5.024	6.635	7.879	1
2	0.010	0.020	0.051	0.103	0.211	4.605	5.991	7.378	9.210	10.597	2
3	0.072	0.115	0.216	0.352	0.584	6.251	7.815	9.348	11.345	12.838	3
4	0.207	0.297	0.484	0.711	1.064	7.779	9.488	11.143	13.277	14.860	4
5	0.412	0.554	0.831	1.145	1.610	9.236	11.070	12.833	15.086	16.750	5
6	0.676	0.872	1.237	1.635	2.204	10.645	12.592	14.449	16.812	18.548	6
7	0.989	1.239	1.690	2.167	2.833	12.017	14.067	16.013	18.475	20.278	7
8	1.344	1.646	2.180	2.733	3.490	13.362	15.507	17.535	20.090	21.955	8
9	1.735	2.088	2.700	3.325	4.168	14.684	16.919	19.023	21.666	23.589	9
10	2.156	2.558	3.247	3.940	4.865	15.987	18.307	20.483	23.209	25.188	10
11	2.603	3.053	3.816	4.575	5.578	17.275	19.675	21.920	24.725	26.757	11
12	3.074	3.571	4.404	5.226	6.304	18.549	21.026	23.337	26.217	28.300	12
13	3.565	4.107	5.009	5.892	7.042	19.812	22.362	24.736	27.688	29.819	13
14	4.075	4.660	5.629	6.571	7.790	21.064	23.685	26.119	29.141	31.319	14
15	4.601	5.229	6.262	7.261	8.547	22.307	24.996	27.488	30.578	32.801	15
16	5.142	5.812	6.908	7.962	9.312	23.542	26.296	28.845	32.000	34.267	16
17	5.697	6.408	7.564	8.672	10.085	24.769	27.587	30.191	33.409	35.718	17
18	6.265	7.015	8.231	9.390	10.865	25.989	28.869	31.526	34.805	37.156	18
19	6.844	7.633	8.907	10.117	11.651	27.204	30.144	32.852	36.191	38.582	19
20	7.434	8.260	9.591	10.851	12.443	28.412	31.410	34.170	37.566	39.997	20
21	8.034	8.897	10.283	11.591	13.240	29.615	32.671	35.479	38.932	41.401	21
22	8.643	9.542	10.982	12.338	14.041	30.813	33.924	36.781	40.289	42.796	22
23	9.260	10.196	11.689	13.091	14.848	32.007	35.172	38.076	41.638	44.181	23
24	9.886	10.856	12.401	13.848	15.659	33.196	36.415	39.364	42.980	45.559	24
25	10.520	11.524	13.120	14.611	16.473	34.382	37.652	40.646	44.314	46.928	25
26	11.160	12.198	13.844	15.379	17.292	35.563	38.885	41.923	45.642	48.290	26
27	11.808	12.879	14.573	16.151	18.114	36.741	40.113	43.195	46.963	49.645	27
28	12.461	13.565	15.308	16.928	18.939	37.916	41.337	44.461	48.278	50.993	28
29	13.121	14.256	16.047	17.708	19.768	39.087	42.557	45.722	49.588	52.336	29
30	13.787	14.953	16.791	18.493	20.599	40.256	43.773	46.979	50.892	53.672	30
31	14.458	15.655	17.539	19.281	21.434	41.422	44.985	48.232	52.191	55.003	31
32	15.134	16.362	18.291	20.072	22.271	42.585	46.194	49.480	53.486	56.328	32
33	15.815	17.074	19.047	20.867	23.110	43.745	47.400	50.725	54.776	57.648	33
34	16.501	17.789	19.806	21.664	23.952	44.903	48.602	51.996	56.061	58.964	34
35	17.192	18.509	20.569	22.465	24.797	46.059	49.802	53.203	57.342	60.275	35
36	17.887	19.223	21.336	23.269	25.643	47.212	50.998	54.437	58.619	61.581	36
37	18.586	19.960	22.106	24.075	26.492	48.363	52.192	55.668	59.892	62.883	37
38	19.289	20.691	22.878	24.884	27.343	49.513	53.384	56.896	61.162	64.181	38
39	19.996	21.426	23.654	25.695	28.196	50.660	54.572	58.120	62.428	65.476	39
40	20.707	22.164	24.433	26.509	29.051	51.805	55.758	59.342	63.691	66.766	40
45	24.311	25.901	28.366	30.612	33.350	57.505	61.656	65.410	69.957	73.166	45
50	27.991	29.707	32.357	34.764	37.689	63.167	67.505	71.420	76.154	79.490	50
55	31.735	33.570	36.398	38.958	42.060	68.796	73.311	77.380	82.292	85.749	55
60	35.534	37.485	40.482	43.188	46.459	74.397	79.082	83.298	88.379	91.952	60
65	39.383	41.444	44.603	47.450	50.883	79.973	84.821	89.177	94.422	98.105	65
70	43.275	45.442	48.758	51.739	55.329	85.527	90.531	95.023	100.425	104.215	70
75	47.206	49.475	52.942	56.054	59.795	91.061	96.217	100.839	106.393	110.286	75
80	51.172	53.540	57.153	60.391	64.278	96.578	101.879	106.629	112.329	116.321	80
85	55.170	57.634	61.389	64.749	68.777	102.079	107.522	112.393	118.236	122.325	85
90	59.196	61.754	65.647	69.126	73.291	107.565	113.145	118.136	124.116	128.299	90
95	63.250	65.898	69.925	73.520	77.818	113.038	118.752	123.858	129.973	134.247	95
100	67.328	70.065	74.222	77.929	82.358	118.498	124.342	129.561	135.807	140.169	100

Answers

1 Discrete probability distributions

EXERCISE 1A

1 **(a)**

x	1	2	3	4
$P(X = x)$	a	$8a$	$27a$	$64a$

(b) $a = \frac{1}{100}$; **(c)** $P(X < 3) = 0.09$.

2 **(a)** $s = 0.14$; **(b)** $P(R \geqslant 8) = 0.81$.

3 **(a)** $c = 0.1$; **(b)** **(i)** $P(Y < 0) = 0.2$, **(ii)** $P(-1 < Y \leqslant 2) = 0.5$.

EXERCISE 1B

1 $E(X) = 7.1$.

2 **(a)**

x	0	1	2
$P(X = x)$	$\frac{1}{6}$	$\frac{5}{9}$	$\frac{5}{18}$

(b) $E(X) = 1\frac{1}{9}$.

3 **(a)**

s	0	1	2	3
$P(S = s)$	0.512	0.384	0.096	0.008

(b) $S \sim B(3, 0.2) \Rightarrow \mu = E(S) = 3 \times 0.2 = 0.6$; **(c)** $P(S \leqslant 1) = 0.896$.

4 **(a)** $k = \frac{1}{90}$; **(b)** $E(R) = 6.71$.

EXERCISE 1C

1 **(a)**

y^{-2}	0.25	1	1	0.25	0.04	0.01
$P(Y = y)$	0.1	0.2	0.3	0.2	0.1	0.1

$\Rightarrow$

y^{-2}	0.01	0.04	0.25	1
$P(Y = y)$	0.1	0.1	0.3	0.5

(b) **(i)** $E(Y^{-2}) = 0.58$, **(ii)** $E(100Y^{-2}) = 58$.

2 **(a)** **(i)** $E(X) = 3.3$, **(ii)** $E(X^2) = 18.9$; **(b)** **(i)** $E(5X) = 16.5$, **(ii)** $E(25X^2) = 472.5$.

3 **(a)** $E(R^{-1}) = 0.448$; **(b)** $E(P) = 8.96$.

EXERCISE 1D

1 Mean $= \mu = 1$, variance $= \sigma^2 = 1$, standard deviation $= \sigma = 1$.

2 $\mu = E(Y) = 2.3$, $\sigma^2 = \text{Var}(Y) = 8.41$, $\sigma = \sqrt{\text{Var}(Y)} = 2.9$.

3 **(a)** $p = 0.1$; **(b)** $\mu = E(T) = 1.5$, $\sigma^2 = \text{Var}(T) = 1.45$, $\sigma = \sqrt{\text{Var}(T)} = 1.20$.

4 **(a)** $\mu = E(X) = 2$, $\sigma = \sqrt{\text{Var}(X)} = \sqrt{2} = 1.41$.

EXERCISE IE

1 **(a)** $\mu = E(X) = 2.9$, $\sigma^2 = \text{Var}(X) = 1.19$;   **(b)** **(i)** 8.7,   **(ii)** 10.71.

2 **(a)** $E(Y) = 4$, $\text{Var}(Y) = 4$;   **(b)** **(i)** 44,   **(ii)** 400.

3 **(a)** Use of tree diagram;   **(b)** $E(R) = 1.2$, $\text{Var}(R) = \dfrac{32}{75}$;   **(c)** $10\frac{2}{3}$.

4 **(a)** $Y \sim B(10, 0.4)$;   **(b)** $E(Y) = 4$, $\sigma = $ standard deviation of $Y = \sqrt{2.4} = 1.549$.   **(c)** **(i)** 48,   **(ii)** 240.

5 **(a)** $E(Y) = 28$, $\sigma = \sqrt{\text{Var}(Y)} = \sqrt{12.6} = 3.55$;   **(b)** 56, 7.10.

    (c) **(i)** 32,   **(ii)** £55 200.

    (iii) By including $Y = 0$, the probability distribution becomes more spread out and so the standard deviation will increase, whilst the mean will decrease.

MIXED EXERCISE

1 **(a)** $k = 0.45$;   **(b)** mean $= \mu = 2.2$, variance $= \sigma^2 = 0.76$, standard deviation $= \sigma = 0.872$.

2 mean $= \mu = 5.7$, variance $= \sigma^2 = 4.27$, standard deviation $= \sigma = 2.07$.

3 mean $= \mu = 1.7$, standard deviation $= \sigma = 1.18$.

4 **(a)**

$\dfrac{1}{r}$	1	0.5	0.25	0.2
P($R = r$)	0.1	0.3	0.2	0.4

    (b) $E(R^{-1}) = 0.38$ and $E(R^{-2}) = 0.2035$;

    (c) $\text{Var}\left(\dfrac{10}{R}\right) = 5.91$ and standard deviation $= 2.431$.

5 **(a)**

x	2	4	6	8
P($X = x$)	0.1	0.2	0.3	0.4

    (b) mean $= \mu = E(X) = 6$ and $\sigma^2 = \text{Var}(X) = 4$;   **(c)** $\mu = E(P) = 70$ and $\sigma(P) = 20$.

6 **(a)** $E(X) = 68.8$ and $\sigma = 21.6$;   **(b)** 0.0464.

7 **(a)** $\mu = 4$ and $\sigma = \sqrt{8} = 2.828$;   **(b)**

x	0	5
P($X = x$)	0.5	0.5

8 **(a)** $\mu = 1.34$ and $\sigma = 1.834$;

    (b) **(i)** $Y \sim B(5, 0.25)$,   **(ii)** $\mu = np = 1.25$ and $\sigma = \sqrt{np(1 - p)} = \sqrt{0.9375} = 0.968$.

    (c) Means are fairly similar. Standard deviations are very different which suggests that the applicants are **not** guessing and that the distribution for Y is **not** binomial.

9 **(a)** **(i)** $E(X) = 1.83$,   **(ii)** $E(X^2) = 5.35$,   **(iii)** standard deviation $= 1.415$;

    (b) **(i)** $P(X < 3) = 0.67$,   **(ii)** $P(X > \mu) = 0.53$.

10 **(a)** $P(X \geqslant 2) = 0.55$;

    (b) **(i)** $E(X) = 1.84$,   **(ii)** $E(X^2) = 4.9$,   **(iii)** $\sigma = 1.23$.

11 **(a)** **(i)** $\mu = 1.47$,   **(ii)** $E(X^2) = 3.37$,   **(iii)** $\sigma = 1.10$;

    (b) **(i)** Binomial $(n = 5, p = 0.4)$,   **(ii)** $\mu = 2$ and $\sigma = 1.10$.

12 **(a)** $E[5(R - 1)] = 10$ and $\text{Var}[5(R - 1)] = 25$;   **(b)** $E(12R^{-1}) = 4.8$ and $\text{Var}(12R^{-1}) = 6.96$.

13 **(b)** $E(C) = 100$ and $\text{Var}(C) = 1035$; **(c)** $E(R) = 383.5$.

14 **(a)** $E(P) = 40$ and $\text{Var}(P) = 900$; **(b)** **(ii)** $E(A) = 150$.

15 **(b)** $E(P) = 10$ and $\text{Var}(P) = 10.8$; **(c)** **(i)** $C = 196 - 4R$, **(ii)** $E(C) = 188$ and $\text{Var}(C) = 19.2$.

16 **(a)** $E[5(2R - 1)] = 25$ and $\text{Var}[5(2R - 1)] = 120$;

(b) **(i)**

$\dfrac{60}{R}$	60	30	20	15	12
$P(R = r)$	0.1	0.2	0.4	0.2	0.1

(iii) $\text{Var}\left(\dfrac{60}{R}\right) = 173.76$.

2 The Poisson distribution

Answers have been given to four decimal places. However to three significant figures is sufficient.

EXERCISE 2A

1 **(a)** 0.8472; **(b)** 0.8488; **(c)** 0.4457; **(d)** 0.5928; **(e)** 0.1377.

2 **(a)** 0.2381; **(b)** 0.8893; **(c)** 0.1954; **(d)** 0.7108; **(e)** 0.3712.

3 **(a)** 0.1048; **(b)** 0.0895; **(c)** 0.1013; **(d)** 0.6894; **(e)** 0.7720.

4 **(a)** 0.0403; **(b)** 0.7479; **(c)** 0.7313; **(d)** 0.1067; **(e)** 0.2480.

5 **(a)** 0.4628; **(b)** 0.1607.

6 **(a)** 0.6728; **(b)** 0.1463.

7 **(a)** 0.8335; **(b)** 0.0394; **(c)** 0.4082.

EXERCISE 2B

1 3.

2 **(a)** 13; **(b)** 15; **(c)** 17.

3 **(a)** 17; **(b)** 21.

4 **(a)** 20; **(b)** 23; **(c)** 24.

EXERCISE 2C

1 **(a)** 0.004 99; **(b)** 0.0265; **(c)** 0.0701; **(d)** 0.124; **(e)** 0.775; **(f)** 0.220.

2 **(a)** 0.0523; **(b)** 0.154; **(c)** 0.228; **(d)** 0.224; **(e)** 0.793; **(f)** 0.606.

3 **(a)** 0.432; **(b)** 0.363; **(c)** 0.152; **(d)** 0.0426; **(e)** 0.0533; **(f)** 0.558.

EXERCISE 2D

1 **(a)** 0.3374; **(b)** 0.2560; **(c)** 0.1377; **(d)** 0.5928.

2 **(a)** 0.1665; **(b)** 0.0985; **(c)** 0.6575; **(d)** 0.0479; **(e)** 0.1757.

3 **(a)** 0.9921; **(b)** 0.1429; **(c)** 0.6512; **(d)** 0.2424.

4 **(a)** 0.9909; **(b)** 0.5940; **(c)** 0.6919; **(d)** 0.2224; **(e)** 0.0651.

5 **(a)** 0.2090; **(b)** 0.1378; **(c)** 0.3712; **(d)** 0.7668; **(e)** 0.8444; **(f)** 0.4075.

EXERCISE 2E

(a) Yes, Poisson likely to be valid;

(b) Where there is congestion lorries won't 'flow' independently so Poisson won't be valid;

(c) Unlikely as cars won't pass the point independently;

(d) Not Poisson, since there is an upper limit on the number of components;

(e) Yes, Poisson likely to be valid;

(f) Likely to be Poisson (although near Christmas the average rate is likely to change);

(g) Poisson likely;

(h) Not Poisson, mean not constant;

(i) Poisson likely;

(j) Poisson likely;

(k) If more than one person injured in an accident, injuries not independent, probably not Poisson.

MIXED EXERCISE

1 **(a)** 0.1653;   **(b)** 0.2694.

2 **(a)** 0.7787;   **(b)** 0.1254.

3 0.2213.

4 **(a)** **(i)** 0.5768,   **(ii)** 0.9881   **(iii)** 0.0116;   **(b)** 18.

5 **(a)** 0.2149;   **(b)** 0.0424; Rate 1.6 per hour.

6 **(a)** The rate of arrival will vary – probably higher near 'mealtimes', lower at night. Not Poisson;

   (b) Not Poisson, calls can't arrive independently and at random;

   (c) Not Poisson since there is an upper limit of 150.

7 **(a)** **(i)** 0.3027,   **(ii)** 0.6665,   **(iii)** 0.0273;

   (b) 0.0290;

   (c) Demand may vary its rate with the seasons or for different days of the week.

8 **(a)** 0.5305;   **(b)** The average rate will vary e.g. more on birthdays or near Christmas.

9 **(a)** **(i)** 0.7306,   **(ii)** 5;

   (b) **(i)** Poisson. Conditions fulfilled.
        (ii) Not Poisson. Calls interfere with each other. Not random.
        (iii) Not Poisson. Mean not constant.

10 **(a)** **(i)** 0.8335,   **(ii)** 0.3081;

   (b) 0.7787;   **(c)** 0.0909;   **(d)** mean may vary according to day of week e.g. more on Saturday.

11 **(a)** **(i)** 0.3012,   **(ii)** 0.2169;

   (b) 0.0160;   **(c)** 0.8666;   **(d)** not independent – some cars will contain more than one passenger.

12 **(a)** 1.90;   **(b)** **(i)** 0.6973,   **(ii)** 0.339;

   (c) $e^{-7.2} \approx 0.000\,746\,5$ ($= 0.0007$ to 4 d.p.)

3 Continuous probability distributions

1 (a) $k = \dfrac{2}{15}$; **(b)** 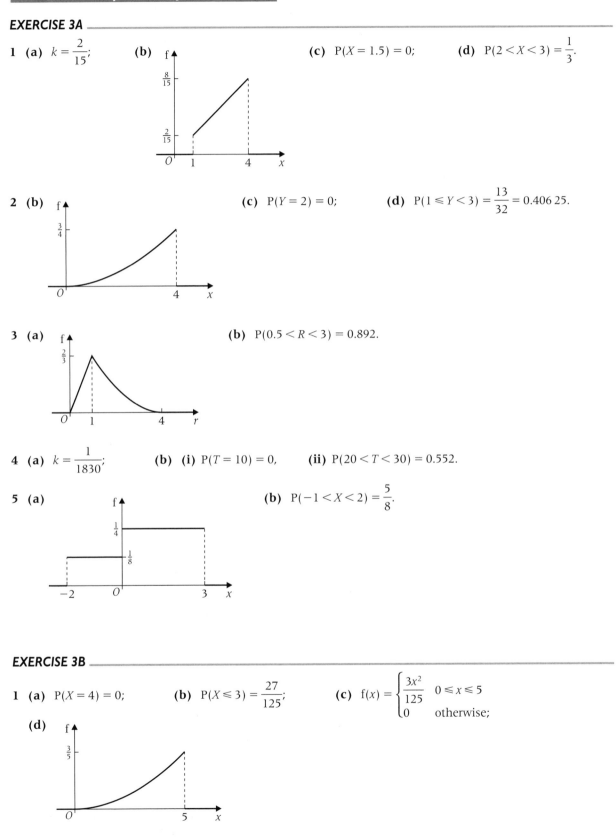 **(c)** P$(X = 1.5) = 0$; **(d)** P$(2 < X < 3) = \dfrac{1}{3}$.

2 (b) **(c)** P$(Y = 2) = 0$; **(d)** P$(1 \leqslant Y < 3) = \dfrac{13}{32} = 0.406\,25$.

3 (a) **(b)** P$(0.5 < R < 3) = 0.892$.

4 (a) $k = \dfrac{1}{1830}$; **(b) (i)** P$(T = 10) = 0$, **(ii)** P$(20 < T < 30) = 0.552$.

5 (a) **(b)** P$(-1 < X < 2) = \dfrac{5}{8}$.

1 (a) P$(X = 4) = 0$; **(b)** P$(X \leqslant 3) = \dfrac{27}{125}$; **(c)** $\mathrm{f}(x) = \begin{cases} \dfrac{3x^2}{125} & 0 \leqslant x \leqslant 5 \\ 0 & \text{otherwise;} \end{cases}$

(d)

2 (a)

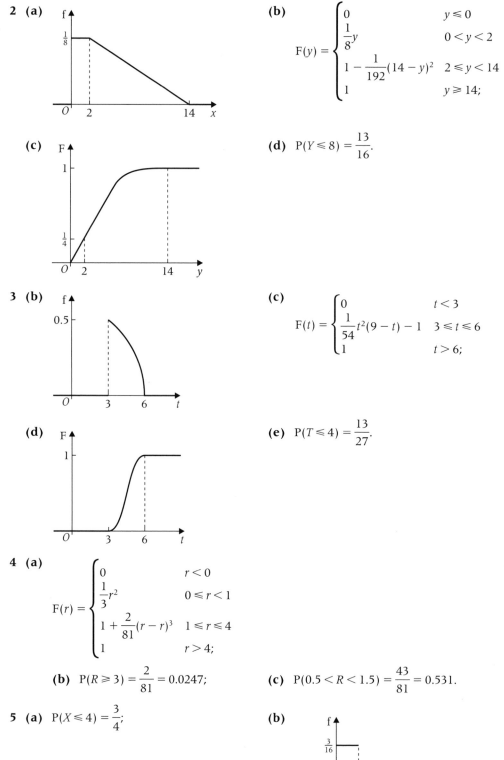

(b)

$$F(y) = \begin{cases} 0 & y \leqslant 0 \\ \dfrac{1}{8}y & 0 < y < 2 \\ 1 - \dfrac{1}{192}(14 - y)^2 & 2 \leqslant y < 14 \\ 1 & y \geqslant 14; \end{cases}$$

(c)

(d) $P(Y \leqslant 8) = \dfrac{13}{16}.$

3 (b)

(c)

$$F(t) = \begin{cases} 0 & t < 3 \\ \dfrac{1}{54}t^2(9 - t) - 1 & 3 \leqslant t \leqslant 6 \\ 1 & t > 6; \end{cases}$$

(d)

(e) $P(T \leqslant 4) = \dfrac{13}{27}.$

4 (a)

$$F(r) = \begin{cases} 0 & r < 0 \\ \dfrac{1}{3}r^2 & 0 \leqslant r < 1 \\ 1 + \dfrac{2}{81}(r - r)^3 & 1 \leqslant r \leqslant 4 \\ 1 & r > 4; \end{cases}$$

(b) $P(R \geqslant 3) = \dfrac{2}{81} = 0.0247;$

(c) $P(0.5 < R < 1.5) = \dfrac{43}{81} = 0.531.$

5 (a) $P(X \leqslant 4) = \dfrac{3}{4};$

(b)

EXERCISE 3C

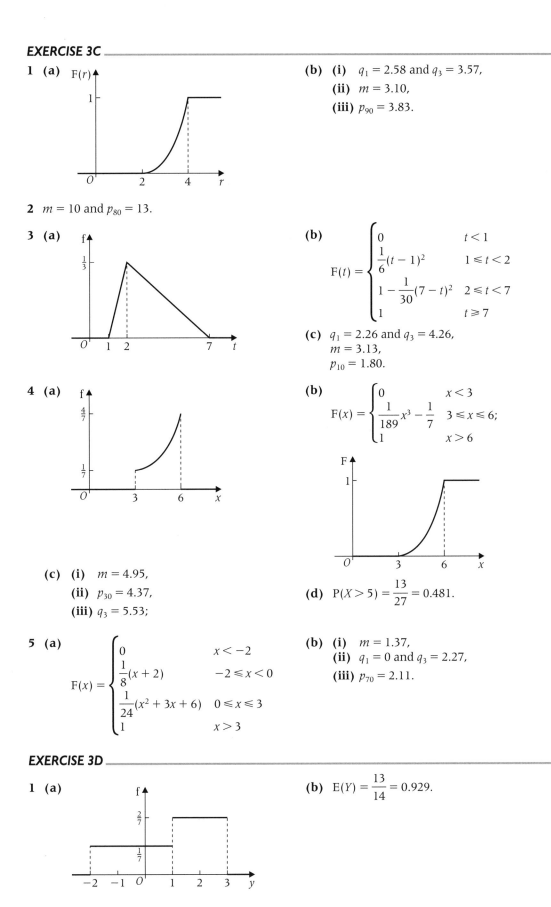

1 (a)

(b) (i) $q_1 = 2.58$ and $q_3 = 3.57$,

   (ii) $m = 3.10$,

   (iii) $p_{90} = 3.83$.

2 $m = 10$ and $p_{80} = 13$.

3 (a)

(b)
$$F(t) = \begin{cases} 0 & t < 1 \\ \dfrac{1}{6}(t-1)^2 & 1 \le t < 2 \\ 1 - \dfrac{1}{30}(7-t)^2 & 2 \le t < 7 \\ 1 & t \ge 7 \end{cases}$$

(c) $q_1 = 2.26$ and $q_3 = 4.26$,
   $m = 3.13$,
   $p_{10} = 1.80$.

4 (a)

(b)
$$F(x) = \begin{cases} 0 & x < 3 \\ \dfrac{1}{189}x^3 - \dfrac{1}{7} & 3 \le x \le 6; \\ 1 & x > 6 \end{cases}$$

(c) (i) $m = 4.95$,

   (ii) $p_{30} = 4.37$,

   (iii) $q_3 = 5.53$;

(d) $P(X > 5) = \dfrac{13}{27} = 0.481$.

5 (a)
$$F(x) = \begin{cases} 0 & x < -2 \\ \dfrac{1}{8}(x+2) & -2 \le x < 0 \\ \dfrac{1}{24}(x^2 + 3x + 6) & 0 \le x \le 3 \\ 1 & x > 3 \end{cases}$$

(b) (i) $m = 1.37$,
   (ii) $q_1 = 0$ and $q_3 = 2.27$,
   (iii) $p_{70} = 2.11$.

EXERCISE 3D

1 (a)

(b) $E(Y) = \dfrac{13}{14} = 0.929$.

2 (a)

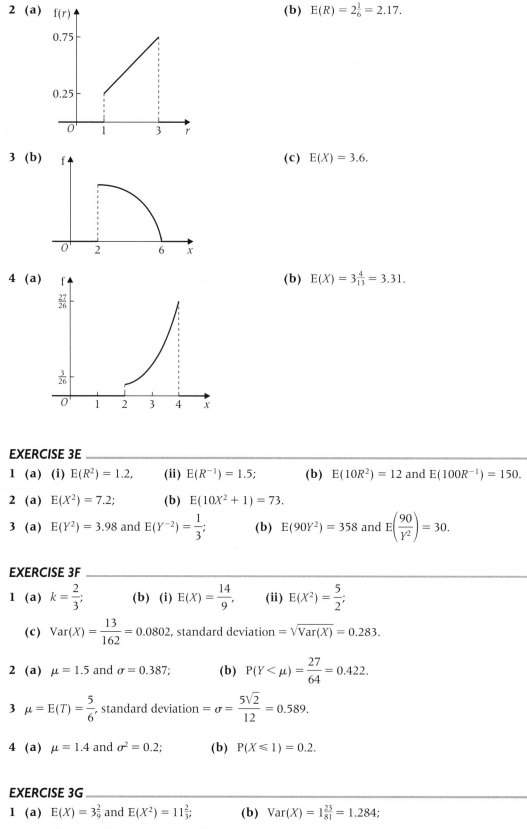

(b) $E(R) = 2\frac{1}{6} = 2.17$.

3 (b)

(c) $E(X) = 3.6$.

4 (a)

(b) $E(X) = 3\frac{4}{13} = 3.31$.

EXERCISE 3E

1 (a) (i) $E(R^2) = 1.2$, **(ii)** $E(R^{-1}) = 1.5$; **(b)** $E(10R^2) = 12$ and $E(100R^{-1}) = 150$.

2 (a) $E(X^2) = 7.2$; **(b)** $E(10X^2 + 1) = 73$.

3 (a) $E(Y^2) = 3.98$ and $E(Y^{-2}) = \dfrac{1}{3}$; **(b)** $E(90Y^2) = 358$ and $E\left(\dfrac{90}{Y^2}\right) = 30$.

EXERCISE 3F

1 (a) $k = \dfrac{2}{3}$; **(b) (i)** $E(X) = \dfrac{14}{9}$, **(ii)** $E(X^2) = \dfrac{5}{2}$;

   (c) $\mathrm{Var}(X) = \dfrac{13}{162} = 0.0802$, standard deviation $= \sqrt{\mathrm{Var}(X)} = 0.283$.

2 (a) $\mu = 1.5$ and $\sigma = 0.387$; **(b)** $P(Y < \mu) = \dfrac{27}{64} = 0.422$.

3 $\mu = E(T) = \dfrac{5}{6}$, standard deviation $= \sigma = \dfrac{5\sqrt{2}}{12} = 0.589$.

4 (a) $\mu = 1.4$ and $\sigma^2 = 0.2$; **(b)** $P(X \leqslant 1) = 0.2$.

EXERCISE 3G

1 (a) $E(X) = 3\frac{2}{9}$ and $E(X^2) = 11\frac{2}{3}$; **(b)** $\mathrm{Var}(X) = 1\frac{23}{81} = 1.284$;

   (c) (i) $\mathrm{Var}(10X) = 128.4$, **(ii)** $\mathrm{Var}(10X - 7) = 128.4$.

2 (a) $E\left(\dfrac{1}{X^2}\right) = \dfrac{1}{40}$;   **(b)** $\text{Var}\left(\dfrac{1}{X}\right) = 0.000\,802$.

3 (b) $E(R^{-1}) = \dfrac{75}{124} = 0.605$ and $E(R^{-2}) = \dfrac{35}{93} = 0.376$;

    (c) (i) $\text{Var}(R^{-1}) = 0.0105$,   **(ii)** $\text{Var}(10R^{-1}) = 1.05$.

4 (a)

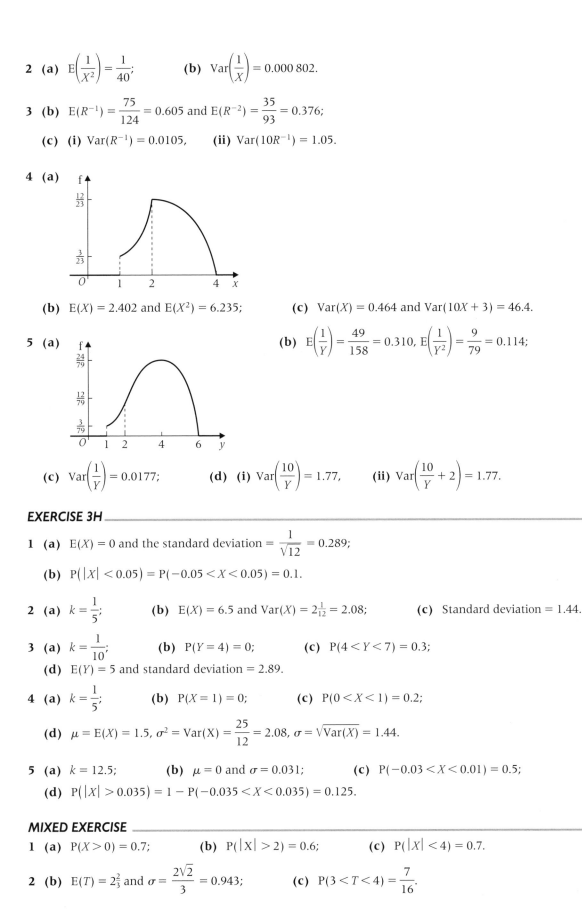

    (b) $E(X) = 2.402$ and $E(X^2) = 6.235$;   **(c)** $\text{Var}(X) = 0.464$ and $\text{Var}(10X + 3) = 46.4$.

5 (a)   **(b)** $E\left(\dfrac{1}{Y}\right) = \dfrac{49}{158} = 0.310$, $E\left(\dfrac{1}{Y^2}\right) = \dfrac{9}{79} = 0.114$;

    (c) $\text{Var}\left(\dfrac{1}{Y}\right) = 0.0177$;   **(d) (i)** $\text{Var}\left(\dfrac{10}{Y}\right) = 1.77$,   **(ii)** $\text{Var}\left(\dfrac{10}{Y} + 2\right) = 1.77$.

EXERCISE 3H

1 (a) $E(X) = 0$ and the standard deviation $= \dfrac{1}{\sqrt{12}} = 0.289$;

    (b) $P(|X| < 0.05) = P(-0.05 < X < 0.05) = 0.1$.

2 (a) $k = \dfrac{1}{5}$;   **(b)** $E(X) = 6.5$ and $\text{Var}(X) = 2\tfrac{1}{12} = 2.08$;   **(c)** Standard deviation $= 1.44$.

3 (a) $k = \dfrac{1}{10}$;   **(b)** $P(Y = 4) = 0$;   **(c)** $P(4 < Y < 7) = 0.3$;

    (d) $E(Y) = 5$ and standard deviation $= 2.89$.

4 (a) $k = \dfrac{1}{5}$;   **(b)** $P(X = 1) = 0$;   **(c)** $P(0 < X < 1) = 0.2$;

    (d) $\mu = E(X) = 1.5$, $\sigma^2 = \text{Var(X)} = \dfrac{25}{12} = 2.08$, $\sigma = \sqrt{\text{Var}(X)} = 1.44$.

5 (a) $k = 12.5$;   **(b)** $\mu = 0$ and $\sigma = 0.031$;   **(c)** $P(-0.03 < X < 0.01) = 0.5$;

    (d) $P(|X| > 0.035) = 1 - P(-0.035 < X < 0.035) = 0.125$.

MIXED EXERCISE

1 (a) $P(X > 0) = 0.7$;   **(b)** $P(|X| > 2) = 0.6$;   **(c)** $P(|X| < 4) = 0.7$.

2 (b) $E(T) = 2\tfrac{2}{3}$ and $\sigma = \dfrac{2\sqrt{2}}{3} = 0.943$;   **(c)** $P(3 < T < 4) = \dfrac{7}{16}$.

3 (a) (ii) median = 5.45;   **(b)** $\mu = E(T) = 5.325$.

4 (b) $E(Y) = 4$;   **(c) (i)** $E(Y^2) = 49\frac{1}{3}$,   **(ii)** standard deviation $= \sqrt{\text{Var}(Y)} = 5.77$;
(d) (i) $P(Y < 0) = 0.3$,   **(ii)** $P(-2 < Y < 2) = 0.2$,   **(iii)** $P(|Y| > 4) = 0.6$.

5 (b) (i) $P(X > 20) = 0.25$,   **(ii)** $P(X < 10) = 0.6$.

6 (b) $P(T < 15) = 0.7$.

7 (a)

   (b) (i) $P(3 \leqslant T < 6) = 0.5$,   **(ii)** $P(T > 6) = \dfrac{1}{3}$,   **(iii)** lower quartile $q_1 = 3.5$;   **(c)** $\dfrac{4}{81}$.

8 (a)

   (c) (i) 0.28,   **(ii)** 0.6.

9 (b) $E(X) = \dfrac{1}{3}p + \dfrac{1}{2}q$;   **(c)** $q = 0.4$;   **(d)** $\sigma = 0.271$.

10 (b) (i) $P(5 < X < 20) = \dfrac{4}{9}$,   **(ii)** $P\left(X < \mu - \dfrac{\sigma\sqrt{3}}{2}\right) = P(X < 16.5) = 0.25$.

11 (b) $\mu = \dfrac{4}{3}$ and $\sigma = \dfrac{1}{3}\sqrt{2} = 0.4714$;   **(c)** $P(T < 1) = 0.25$.

12 (b) $\mu = E(X) = 1.25$ and $\sigma = 0.829$.

13 (a)

   (c) $P(H > 2) = \dfrac{1}{3}$;

14 (b) $P(S < 3) = 0.210$;   **(c)** median $= \sqrt[3]{63} = 3.98 \simeq 4$ minutes.

15 (a) $P(X > 6) = 0.352$;   **(b)** $E(X) = 5$.

4 Confidence intervals

EXERCISE 4A

1 (a) 93.50–98.50;   (b) Operating time reduced – overhaul effective.

2 (a) 220.69–224.65;   (b) 220.06–225.28;   (c) 219.47–225.86;   (d) 218.15–227.18.

3 (a) (i) 2347.5–2399.4,   (ii) 2329.6–2417.2;
   (b) First confidence interval uses data from past experience. Likely to be more accurate unless standard deviation has changed. Second confidence is valid whether or not standard deviation is 35.

4 (a) (i) 510.09–514.25,   (ii) 508.69–515.64;
   (b) First confidence interval uses data from past experience. Likely to be more accurate unless standard deviation has changed. Second confidence is valid whether or not standard deviation is 2.6.

5 (a) 24.52–29.64;   (b) 18.85;
   (c) 7.40   Regulations appear to be adequate as 7.40 is well below the load at which an anchor might fail.

6 (a) 18.9–63.1;   (b) Sample random and population normally distributed;
   (c) 32.5–40.1;
   (d) Still need to assume sample random. Normal distribution not necessary as sample large.

7 (a) 454.19–459.31;
   (b) Evidence that mean is above 454 g, but some individual jars contain less than 454 g;
   (c) 0.05.

8 (a) 169–1085;
   (b) (i) distribution skew (3 zeros),   (ii) Negative values impossible;
   (c) (i) 926.0–994.0,
        (ii) Sample large – mean approximately normally distributed by Central Limit Theorem,
        (iii) Consumption on night-shift may differ from consumption on other shifts. Consumption may change if scheme introduced on permanent basis.

9 (a) (i) 490.0–502.2,   (ii) 492.5–499.8;
   (b) 498.0–501.2;
   (c) (i) (A) 0.8, (B) 0.15,   (ii) 0.19.

10 (a) 29.8–67.0;
    (b) (i) 46.3–52.7,   (ii) 17.2–81.8;
    (c) Evidence mean > 25 but some individual scores < 25;
    (d) Interval narrower. No need to assume normal distribution;
    (e) Suggestion is valid since both samples random but sample at 120 only slightly better than 110.

5 Hypothesis testing

EXERCISE 5A

1 ts 1.80   cv 1.6449   mean greater than 135 kg.

2 ts −1.22   cv ±1.96   accept mean reaction time 7.5 s.

3 ts 1.05   cv 1.6449   accept children don't take longer.

4 ts −3.48   cv ±2.5758   mean length not satisfactory (less than 2 cm).

5 ts 1.67   cv 1.6449   silver not pure.

6 ts 3.19   cv 2.3263   mean weight has increased.

7 ts −1.57   cv −1.6449   accept mean weight not reduced.

8 ts 3.95   cv ±1.96   mean time not (greater than) 4 s.

EXERCISE 5B

1 **(a)** **(i)** Conclude mean greater than 135 kg when in fact mean equals 135 kg,
   (ii) Conclude mean equals 135 kg when in fact mean greater than 135 kg;

   (b) **(i)** 0.05,   **(ii)** 0.

2 **(a)** **(i)** Conclude mean unsatisfactory when it is satisfactory,
   (ii) Conclude mean is satisfactory when it is unsatisfactory;

   (b) **(i)** 0.01,   **(ii)** 0.

3 **(a)** ts −1.80   cv ±1.96   accept mean time is 20 minutes;

   (b) Conclude mean time not 20 min when in fact it is 20 minutes;

   (c) would need to know the actual value of the mean.

4 **(a)** ts −1.69   cv ±1.6449   mean length not (less than) 19.25 mm;

   (b) **(i)** Conclude mean not 19.25 mm when in fact it is 19.25 mm,
   (ii) Conclude mean 19.25 mm when in fact it is not 19.25 mm;

   (c) 0.1.

5 **(a)** ts −1.90   cv −1.6449   easier to assemble;

   (b) Conclude not easier to assemble when in fact it is.

6 Further hypothesis testing for means

EXERCISE 6A

1 ts 2.42   cv 2.3263   mean breaking strength greater than 195 kg.

2 ts 1.02   cv ±1.96   mean resistance 1.5.

3 ts −2.67   cv −1.6449   mean test score lower.

4 ts 1.79   cv 1.6449   mean weight greater than 500 g.

5 ts −2.19   cv −2.3263   no significant evidence that suspicions are correct.
Sample assumed random.

6 ts 1.87   cv 2.3263   mean weight not greater than 35 g.

7 (a) ts −4.98   cv −1.6449   mean less than 7.4;

 (b) Test statistic is −4.98 so conclusion would have been unchanged even if the significance level had been very much less than 5%. Conclusion is very clear;

 (c) As the sample is large it is not necessary to make any assumption about the distribution. However it is necessary to assume the sample is random. There is no information on this. If, for example, only those with particular symptoms had been tested they might be untypical of all varicoceles sufferers.

EXERCISE 6B

1 ts −0.532   cv ±2.179   mean length 50 cm.

2 ts 1.43   cv 1.383   mean exceeds 25 kg.

3 ts 2.78   cv 1.771   mean greater than 120 kg.

4 ts −0.937   cv −1.761   mean time not less than 2 minutes.

5 ts −2.36   cv ±2.228   mean not equal to (less than) 95
assumed distribution normal.

6 ts −2.45   cv −2.896   new mean not smaller.

MIXED EXERCISE

1 ts −2.27   cv ±1.96   readings biased.

2 (a) ts −1.35   cv ±1.734   mean diameter 275 mm;

 (b) Use $\sigma = 5$, ts −1.74, cv ±1.6449 from normal tables, conclude mean diameter not equal to (less than) 275 mm.

3 ts 12.4   cv ±2.5758   mean dust deposit not equal to (greater than) 60 g.

4 ts −2.11   cv −1.6449   mean score for new ice cream lower.

5 ts 1.48   cv ±1.812   mean mark 40.

6 ts 2.14   cv 1.6449   mean higher than 77.4 mm.

7 (a) ts 1.49   cv ±2.365   mean 40 minutes;

 (b) Time of three hours was due to exceptional circumstances. If manager wants to examine mean time under normal circumstances it is appropriate to exclude this one. If manager wishes to examine mean time under all circumstances the time of three hours should be considered. If it is included the assumption of normal distribution will be violated.

8 (a) ts 1.91   cv 1.6449   mean life longer;

 (b) use value estimated from sample for standard deviation and obtain critical value from t-distribution;

 (c) ts 1.30   cv 1.833   mean life not longer.
 This conclusion does not mean that we have proved the mean is 960 days only that there is insufficient evidence to disprove it. With the additional information that the standard deviation was 135 days there was sufficient evidence to show that the mean was more than 960 days subject to a risk of 5% of claiming an increase when no increase exists.

9 **(a)** ts 3.19   cv ±1.96   mean take home pay not equal to (greater than) £140;

  (b) **(i)** large sample so conclusions not affected
      (central limit theorem),

    (ii) conclusion unreliable – for example, whole sample could have been taken from one employer who paid relatively high wages.

10 **(a)** ts 1.54   cv 1.6449   mean mark not higher;

  (b) mean of large sample approximately normally distributed by central limit theorem;

  (c) sample was self selected. Probably children of highly motivated/gullible parents. Conclusion unreliable since sample not random.

7 Contingency tables

In this chapter you may obtain slightly different values of X^2 according to the number of decimal places used for the Es.

EXERCISE 7A

1 $X^2 = 38.7$ cv 9.21   time of day associated with type of birth.

2 $X^2 = 23.8$ cv 9.21   proportion favouring road improvements not independent of area.

3 $X^2 = 6.66$ cv 7.815   accept choice of dish independent of day of week.

4 **(a)** Only women in the main shopping area asked (many other answers possible);

  (b) $X^2 = 17.7$ cv 9.49   respondents' replies not independent of age.

5 **(a)** $X^2 = 17.0$ cv 5.99   proportion transferred not independent of surgeon;

  (b) B had substantially less transferred than expected. Need to know how cases were allocated. It could be, say, that only straightforward operations were undertaken by B.

6 **(a)** $X^2 = 11.4$ cv 9.21   sex and grade not independent;

  (b) Females did better – more than expected obtained high grades and less than expected had low grades;

  (c) **(i)** Need total number of candidates in order to calculate frequencies,

    (ii) cv 10.6   subject and grade not independent,

    (iii) A bigger proportion of statistics candidates than pure and applied candidates obtained high grades. Would need some information on quality and preparation of candidates to say whether it is easier to get a good grade in statistics.

7 $X^2 = 16.9$ cv 13.3
Grade and length of employment not independent. Unskilled staff tend to stay longer than skilled staff. In particular more unskilled employees than expected stay longer than five years and more skilled staff than expected leave within two years.

EXERCISE 7B

1 $X^2 = 12.4$ cv 5.99   age and sex not independent.

2 $X^2 = 3.14$ cv 7.815   (5%) (another % could be chosen as not specified in question) accept choice independent of gender.

3 $X^2 = 1.99$ cv 9.21   accept method of communication independent of subject.

4 (a) $X^2 = 35.3$ cv 7.815   proportion of guests rating a feature important not independent of feature;

(b) Availability of squash courts much less important than other features. Little to choose between other features – number rating them important exceeded expected number by roughly the same proportion;

(c) If $E < 5$;

(d) Comfortable beds – most similar category;

(e) cv between 73.3 and 79.1. Rating and feature not independent.

EXERCISE 7C

1 $X^2 = 5.73$ cv 3.84   proportion exhibiting allergies not independent of treatment. Drug effective – less than expected of those receiving the drug exhibited allergies.

2 $X^2 = 3.50$ cv 3.84   (5%) (another % could be used) accept proportion germinating independent of variety.

3 $X^2 = 0.618$ cv 3.84   accept proportion defective independent of mould.

4 $X^2 = 1.06$ cv 3.84   accept equal proportion of librarians and designers can distinguish the word.

5 (a)

	<87%	>87%
National	10	5
Labour	6	7

(b) $X^2 = 0.507$ cv 3.84   accept winning party independent of turnout;

(c) Constituencies appear to have been selected in alphabetical order. No constituencies from last part of alphabet. This could effect the test but election results unlikely to be associated with name of constituency so test probably valid. (Other answers possible.)

MIXED EXERCISE

1 (a) $X^2 = 8.11$ cv 3.84   reason associated with time of day;

(b) Using crossing because of Beatles more likely out of rush hour.

2 $X^2 = 25.3$ cv 11.3   classification not independent of inspector.

3 (a)

Week	1	2	3	4	5
Statistics tables	24	32	20	18	9
Other items	192	168	146	87	55

(b) $X^2 = 3.56$ cv 9.49   accept proportion of items which were statistical tables independent of week.

4 (a) $X^2 = 7.59$ cv 3.84   colour not independent of habitat;

(b) Greater proportion of woodland snails dark;

(c) Test valid since $Es > 5$;

(d) Test not valid since Os are not frequencies.

5 $X^2 = 18.9$ cv 5.99   returning 1997 questionnaire not independent of answer to 1996 question on truancy;

6 (a) $X^2 = 72.8$ cv 7.815   snoring associated with heart disease;

(b) Snore more e.g. more sufferers snore every night than expected;

(c) No. Heart disease and snoring associated but there is no proof of cause and effect.

7 (a) $X^2 = 33.5$ cv 7.815   outcome not independent of treatment;

(b) New treatment effective since number of those receiving new treatment who show marked improvement greater than expected;

(c)

	Died	Refused	Untraceable
New	19	10	15
Standard	3	12	18

cv 5.99   reason depends on treatment.

(d) New treatment has a high risk of death and so should not be used. This was hidden in first set of data where patients who died were included in the category 'information unobtainable'.

Exam style practice paper

1 (a) $E(X) = 7.0$, $Var(X) = 75.1 - 7^2 = 26.1$;

(b) (i)

x	16	8	4	2	1
$P(X = x)$	0.1	0.15	0.25	0.3	0.2

(ii) $E(T) = 4.6$;

(c) (i) $A = 16 + \dfrac{16}{X}$,   **(ii)** $E(A) = 16 + E(T) = 20.6$.

2 (c) $F(t) = \dfrac{1}{66}(25t - 2t^2 - 12) = 0.5 \Rightarrow t = 2.18$.

3 (a) $H_0 : \mu = 45$, $H_1 : \mu \neq 45$. Under H_0 $\bar{X} \sim N(45, 0.32)$ and
$z_{crit} = \pm 2.5758$
$z = 2.83 > z_{crit} \Rightarrow$ reject H_0.
Evidence to suggest that the mean length of this species of fish has changed (increased).

(b) Type I error : Reject H_0 when H_0 true $\Rightarrow$ suggest mean length of this species of fish has changed when in fact it has not changed at all.

4 H_0 : No association between college and the proportion of students gaining at least one grade A.

H_1 : Proportion of students gaining at least one grade A associated with college.

$\sum \left(\dfrac{|O_i - E_i| - 0.5}{E_i} \right)^2 = 0.935$ and $\chi_{5\%}^2(1) = 3.841 \Rightarrow$ Accept H_0 Insufficient evidence to suggest an association between the college and the proportion of students gaining at least one grade A.

5 (a) $\begin{matrix} H_0 : \mu = 76 \\ H_1 : \mu > 76 \end{matrix} \Rightarrow$ one-tailed test. $\bar{x} = 77.9$ and $s = 11.21$, $t = 0.536$, $t_{1\%}(9) = 2.821$.

Accept H_0 Insufficient evidence to suggest an increase in the mean score thus no real indication that this group of students is any better than in previous years.

(b) Type II error: Accept H_0 when H_0 false $\Rightarrow$ If we had accepted H_0 when it was false, then we would have assumed the present group of students to be no better than those in previous years, when in fact they are.

6 (a) $P(H = 4) = 0.0111$ **or** (0.0112 if tables used);

(b) (i) $X \sim P_0(9.0)$,

(ii) $P(X \geqslant 12) = 1 - P(X \leqslant 11) = 0.197$,

(iii) $p = 0.197^3 = 0.007\ 65$.

7 (a) $\bar{y} = 1.505$ and $S^2 = 0.005\ 371$;

(b) (1.44, 1.57);

(c) 1.40 does not fall within the 95% confidence interval.
There may be several reasons for this value being used in the advertisement. After the garden centre placed the advertisement, the young apple trees may have grown more quickly than predicted. The garden centre were probably a little conservative with their estimate when they indicated this value in their advert.

8 (a) $c = \dfrac{1}{b - a}$;

(c) (i) $\mu = 1$ and $\sigma = \dfrac{4\sqrt{3}}{3} = 2.31$,   (ii) 0.429.

Index

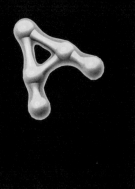